Agricultural and Industrial Microbiology

Agricultural and Industrial Microbiology

By :

Dr. K. Kumutha
Dr. K.G. Anitha
Ms. K. Saranya

SATISH SERIAL PUBLISHING HOUSE
403, Express Tower, Commercial Complex, Azadpur, Delhi-110033 (India)
Phone : 011-27672852, Fax : 91-11-27672046
E-mail :info@satishserial.com, hkjain1975@yahoo.com
Website : www.satishserial.com

Published by :

SATISH SERIAL PUBLISHING HOUSE

403, Express Tower, Commercial Complex, Azadpur, Delhi-110033 (INDIA)
Phone : 011-27672852 Fax : 91-11-27672046
E-mail : info@satishserial.com, hkjain1975@yahoo.com

ISBN 978-93-81226-08-7

Typeset at: Harminder *for* Laxmi Art Creation

Printed at: Salasar Imaging Systems, Delhi-110035

Preface

"Life would not long remain possible in the absence of microbes"
Louis Pasteur

The primary goals of this textbook is to help students to understand and appreciate the natural roles, structure and function of microorganisms and to skyrocket their knowledge in relation to the main branches of Microbiology – *"Agricultural and Industrial Microbiology"*.

In this 21st century Microbiology is emerging as the key biological science as it has a paramount range of application in agriculture and industrial science that yields more economic returns and sounds more wealthy and green. Microbiology is therefore taught to the students all over the world. Agricultural and Industrial Microbiology is taught as one of the important subject to the undergraduate and postgraduate students in Agricultural universities under their usual curriculum. Therefore, in order to benefit the students, this book is written to discuss the basics of microorganisms and give insight coverage of the various aspects of Agricultural and Industrial Microbiology.

The preliminary chapters (Chapters 1 to 11) deal with the general basics of Microbiology like historical development; bacterial morphology and classification; nutrition, growth, control and metabolism of microorganisms. Chapters 12 to 19 cover broadly the basic and applied aspects of Soil Microbiology. For instance, describes microbial interactions, nutrient transformations in soil and biofertilizers.

In the next chapters 20 and 21, a concept of fermentation is introduced along with its history. The chapter 22 briefly explains the types of fermentation. Chapters 23 to 32 discuss the application of microbes in industries to get valuable economic products of high utility like fermented foods, alcoholic beverages, antibiotics, *etc.* In chapter 32 and 34, we discuss the miscellaneous roles of microorganisms in biofuel production and purification of potable water.

The subject matter in this textbook is basic and more information is presented pertaining to 'Agricultural and Industrial Microbiology'. However, books listed in the selected reference section can be read for further deeper understanding of the subject.

We are grateful to all those who have assisted in one way or the other, in bringing out this book. We sincerely hope that this book will help the readers in having better understanding of the basic and applied aspects of Microbiology and suggestions or comments for improving the contents of the text are warmly welcomed.

Final Note of Encouragement

One of life's little truth is that, you get out of any endeavor what you put into it so that the more time you put into it, the more time you spend in serious study, the more you will learn. This will lead to a pride in mastery, greater skill in discovery, and a thrill of learning that is almost like being a Microbiology detective!

Dr. K. Kumutha
Dr. K.G. Anitha
Ms. K. Saranya

Contents

Part-3 : Industrial Microbiology

Part 1

Basic Microbiology

Chapter 1

Introduction

In this universe, when we consider the population of all the organisms, it is noticed that only microorganisms as a group exceeds all others. Microbes are the dominant form of life on planet Earth. More than half the biomass on Earth consists of microorganisms, whereas animals constitute only 15% of the mass of living organisms on Earth. The variety and diversity among microorganisms on earth is no doubt brilliant and important for human life.

Microbiology is about microorganisms and how they work, especially the bacteria, a large group of cells of enormous basic and practical significance. Microbiology deals with microbial diversity and evolution, it envisages information regarding what microorganisms do in the world at large, in human society, in the human body, and in animals and plants. One way or the other, microorganisms affect all other life forms on earth, and thus the science of microbiology is of enormous importance.

Since, microbiology is considered as the foundation for all biological sciences, understanding the basic concepts of microbiology is paramount.

This book deals with

- How and where they live
- Their structure
- How they derive food and energy
- Functions of soil micro flora
- Role in nutrient transformation
- Relation with plant
- Importance in industries
- Microbial products

In this introductory chapter, we introduce the subject of microbiology to provide a general idea of the organisms that microbiologists study.

Microbial world

Microbiology often has been defined as the study of microorganisms and agents too small to be seen clearly by the unaided eye – that is, the study of *microorganisms*. Microorganisms include a large and diverse group of microscopic organisms that exist as single cells or cell clusters, and the viruses, which are microscopic, but not cellular.

One of the first things to realize about the microbial world is that microorganisms are everywhere. Almost every natural surface is colonized by microbes (including your skin). Some microorganisms can live quite happily in boiling hot springs, whereas others form complex microbial communities in frozen sea ice.

As an applied biological science, microbiology deals with many important practical problems in medicine, agriculture and industry. Some of the most important diseases of humans, other animals and plants are caused by microorganisms. Most microorganisms are harmless to humans. You swallow millions of microbes every day with no ill effects. In fact, we are dependent on microbes to help us digest our food and to protect our bodies from pathogens. Microorganisms play major roles in soil fertility and plant productivity. In addition, many large-scale industrial processes, such as the production of antibiotics or human proteins, are microbial based. Microbes also keep the biosphere running by carrying out essential functions such as decomposition of dead animals and plants.

Members of the microbial world

Two structural types of microbial cells are recognized based on the examination of the nucleus structure,

1. Prokaryote
2. Eukaryote

Prokaryotes - The organism with a simple internal structure, lacking true nucleus (membrane enclosed chromosome and nucleolus) and other organelles like mitochondria, golgi body, endoplasmic reticulum etc. are referred as Prokaryotes. Prokaryotes consist of the *Bacteria* and the *Archaea*.

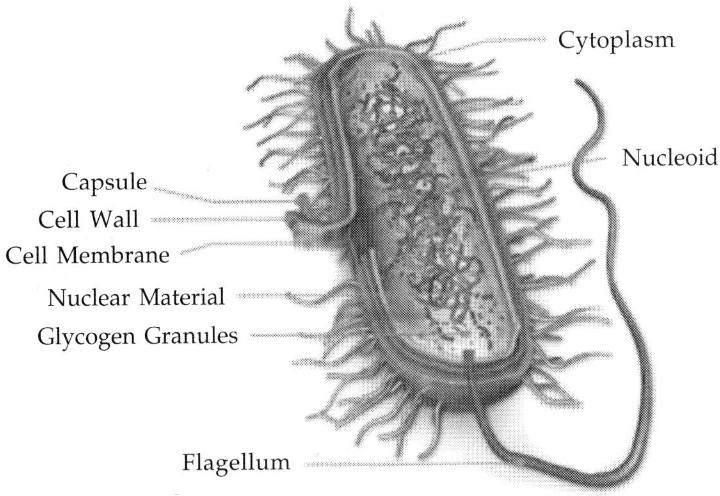

Prokaryotic Cell

Eukaryotes - The organism possessing membrane enclosed nucleus. The major feature of eukaryotic cells, absent in prokaryotic cells, is the presence of membrane-enclosed organelles. Like nucleus, mitochondria, and the chloroplast. Eukaryotic microorganisms include algae, fungi and protozoa. All multicellular plants and animals are constructed of eukaryotic cells.

The major similarities between the two types of cells (prokaryote and eukaryote) are:

- They both have DNA as their genetic material
- They are both membrane bound.
- They both have ribosomes

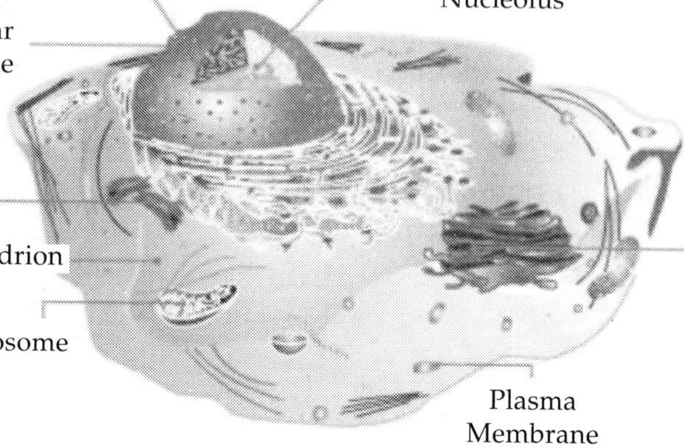

Eukaryotic Cell

- They have similar basic metabolism
- They are both amazingly diverse in forms

Difference between prokaryotic and eukaryotic cells

Characteristics	Prokaryotic cell	Eukaryotic cell
	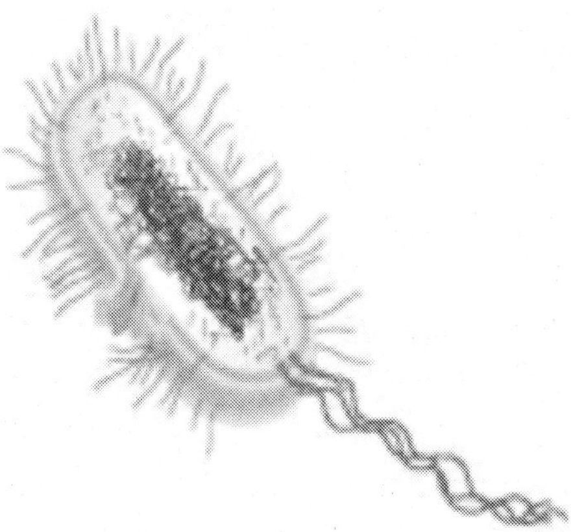	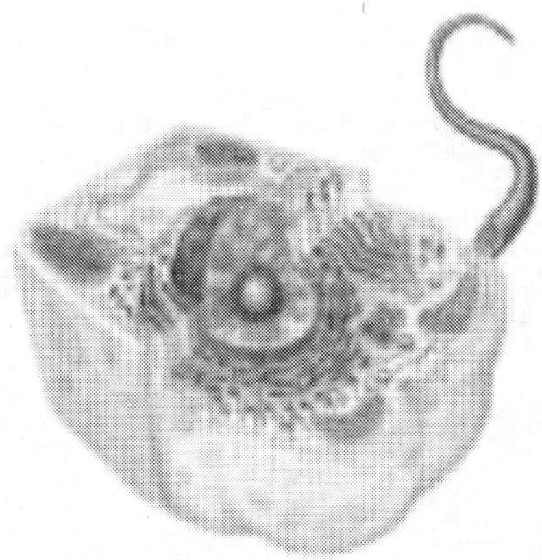
Size of cell	Typically 0.2-2.0 μm in diameter	Typically 10-100 μm in diameter
Nucleus	No nuclear membrane or nucleoli	True nucleus, consisting of nuclear membrane
Membrane-enclosed orgonelles	Absent	Present; examples include lysosomes, Golgi complex, endoplasmic reticulum, mitochondria and chloroplasts

contd...

Flagella	Consist of two protein building blocks	Complex; consist of multiple microtubules
Glycocalyx	Present as a capsule or slime layer	Present in some cells that lock a cell wall
Cell wall	Usually present; chemically complex (typical bacterial cell wall includes peptidoglycan)	When present, chemically simple
Plasma membrane	No carbohydrates and generally lacks sterols	Sterols and carbohydrates that serve as receptors present
Cytoplasm	No cytoskeleton or cytoplasmic streaming	Cytoskeleton; cytoplasmic streaming
Ribosomes	Smaller size (70S)	Larger size (80S); smaller size (70S) in organelles
Chromosome (DNA)	Single circular chromosome; lacks histones	Multiple linear chromosomes with histones arrangement
Cell division	Binary fission	Mitosis
Sexual reproduction	No meiosis; transfer of DNA fragments only	Involves meiosis

Major groups of microorganisms

Based on the phylogenic, morphological and physiological characters, microorganisms were divided into 6 distinct groups. They are as follows:

- Bacteria
- Archaea (or Archaeobacteria)
- Fungi
- Algae
- Protozoa
- Viruses

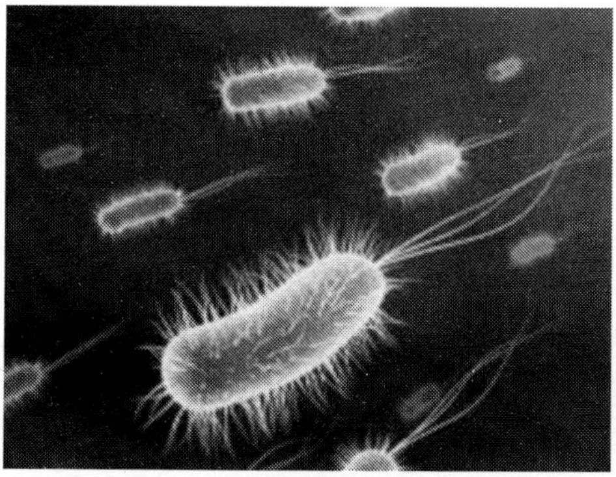
Bacteria

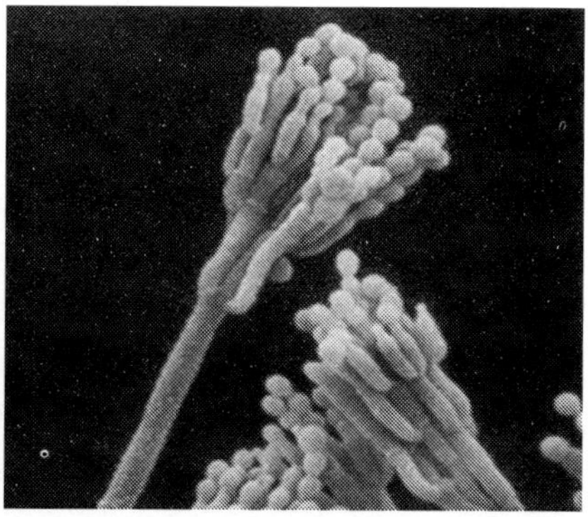
Fungi

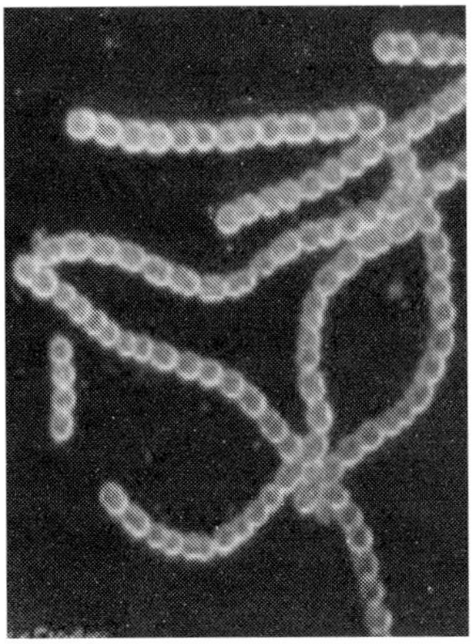

Algae

Protozoa

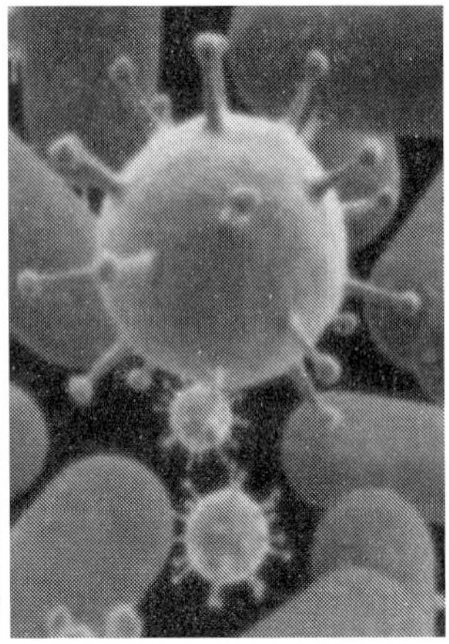

Virus

Bacteria are prokaryotes that are usually single celled organisms. Most have cell walls that contain the structural molecule peptidoglycan. They are abundant in soil, water and air and are also major inhabitants of our skin, mouth and intestines. Some bacteria live in environment that has extreme temperatures, pH or salinity. Although some bacteria, may play more beneficial roles such as cycling elements in the biosphere, breaking down dead plant and animal material and producing vitamins. Cyanobacteria produce significant amounts of oxygen through the process of photosynthesis.

Archaea are prokaryotes that are distinguished from bacteria by many features, most notably their unique ribosomal RNA sequences. They also lack peptidoglycan in their cell walls and have unique membrane lipids. Some have unusual metabolic characteristics, such as the methanogens, which generate methane gas. Many archaea are found in extreme environments.

Fungi are eukaryotic lower plants devoid of chlorophyll. They are usually multicellular but are not differentiated into roots, stems and leaves. They range in size and shape from single-celled microscopic yeasts to gaint multicellular mushrooms. True fungi are composed of filaments and masses of cells which make up the body of the organism, known as mycelium. Fungi reproduce by fission, budding or by means of spores borne on fruiting structures that are quite distinct for certain species.

Algae are relatively simple organisms. The most primitive types are unicellular. Others are aggregations of similar cells with or no differentiation in structure or function. Still other algae have a complex structure with cell types specialized for particular functions. Regardless of size or complexity, all algal cells contain chlorophyll and are capable of photosynthesis. Algae are found most commonly in aquatic environments or in damp soil.

Protozoa are unicellular eukaryotic, animal like protists that are usually motile and lack cell wall. Many free living protozoa function as the primary hunters and grazers of the microbial world. They obtain nutrients by ingesting organic matter and other microbes. They can be found in many different environments and some are normal inhabitants of the intestinal tracts of animals, where they aid in digestion of complex materials such as cellulose. A few cause disease in humans and other animals.

Viruses are group of non-cellular organisms, parasite or pathogen to plant, animals and other microorganisms. They are too small and can be visualized only under electron microscopes.

❑❑❑

Chapter 2

History of Microbiology

Like any science, modern microbiology owes much to its past. Now let us have an insight in to the historical overview of microbiology in this chapter.

Obviously human have to deal with microbes even before the recorded history. Ancient cultures and civilizations had no idea that microbes existed but they did comprehend some of their important effects. For example:

Babylonians were among the earliest peoples, to use yeast in order to make beer (8000 years ago) and acetic acid bacteria to make vinegar (6000 years ago). The Romans had God for that was specific for microorganisms. The roman God of mold and mildew was *"Robigus"* and *"Robigo"* which means crop rust. Ancient Chinese immunized people against smallpox by having them inhale dried, powdered scabs from those suffering from a mild form of the disease. Many traditional cultures have also recognized and used plants as remedies for certain diseases. For example, South Americans recognized the usefulness of extracts of Cinchona tree (containing quinine) to treat malaria. Many cultures also recognized the communicability of certain diseases by microbes in ancient period.

History of microbiology can be divided into 4 periods

1. *Pre historic period (until 1850) :* Period of speculation in nature
2. *Golden era (1850 - 1900) :* Period of fundamental discoveries
3. *Modern era (1900 - 1960) :* Chemotherapy, Antiseptics, Electronic microscope
4. *Molecular era (1960 - till date) :* Research at molecular level

Scientists and their contribution to the development of microbiology

I. Development of microscope

The actual development of microbiology happened only after 1600's, after the discovery of microscope.

Marcus Vavro (116-270BC) - grinding of lenses

Roger Bacon (1220-1292) - clarified the principles of optics and he was credited with being the 'Founder of Optics'.

Galileo Galilei (1564-1642) - invented microscope.

Robert Hooke - He was the builder of the first (or, at least, an early) compound microscope with 30x magnification. He described the fruiting. He coined the term cell and also described fungi but his microscope was unable to resolve bacteria.

Antony Van Leeuwenhoek (1632-1723), was considered by many to be the **"Father of Microbiology"**. Leeuwenhoek was not the first to develop a lens or microscope but was the first to describe living microbes, including bacteria. He constructed 247 complete microscopes with magnifying powers of 40-270 X. He described the "very little animalcules" which we recognize as free-living protozoa and discovered incredible many little animalcules, in a tiny drop of water.

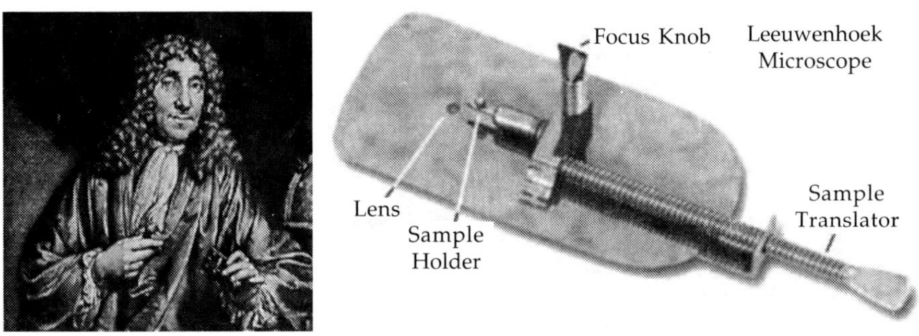

Leeuwenhoek and his microscope that was developed first

Earnest Abbe (1840-1905), perfection in microscopy was obtained by his invention of sub-stage condenser.

II. Conflict over spontaneous theory

Many of the microbiological techniques began to be developed as scientists grabbed with the conflict over the theory of spontaneous generation. This conflict and subsequent studies on the role played by microorganisms in causing disease ultimately led to what is now called as the "**Golden Age of Microbiology**".

From ancient times, people believed in spontaneous generation - **Abiogenesis** (a – not; bio – life; genesis – origin). This long held theory advocated that life could arise spontaneously from nonliving or decomposing matter. For example, many believed that flies could develop directly from rotting meat, wherein many tried to disprove it.

Aristotle (248 to 322BC,) also believed that, life is being originated from non-living things spontaneously from the soil, plants, or other unlike animals.

Francesco Redi (1626-1697), was among the first to question abiogenesis. He placed meat in a jar covered with gauze. Redi provided non microscopic evidence against abiogenesis by showing flies developed from maggots on meat in uncovered jars but not from meat in covered jars. Attracted by the odour of the meat, flies laid eggs on the covering, and from the eggs maggots developed on the gauze.

Redi's experiment

John Needham (1713 to 1781BC), an English clergyman and proponent of abiogenesis, claimed that life arose spontaneously because of a random or chance clumping of "organic" molecules. He observed the appearance of organisms not present at the start of the experiment and concluded that the bacteria originated from the meat.

Spallanzani (1729-1799), tried to refute Needham's work by heating hay broth after it was sealed in a flask preventing microbes from entering by air. Microbes could not grow spontaneously from such reheated broth. His findings were not totally accepted, as many believed that oxygen was vital to sustain life and nothing would grow if there had not been any in flasks.

Louis Pasteur (1822 – 1895), prepared a meat broth in a swan-neck flask and boiled it for several hours. He then left the flask unsealed on a laboratory bench. Previous experiments have shown that mere boiling of the broth did not diminish its generative capacity, so the experiment could not be criticized on these grounds. The flask was not sealed, and there was free exchange of air with the environment, so the system did not lack oxygen. Still, the swan-neck remained free of microbial contamination for months, because their swan-necks were shaped to trap viable microbial particles and to allow only air to enter the flask. After several months when broke the neck of one of these flasks, contamination by air and proliferation of micro-organisms in the fluid ensued. This simple experiment thus altogether disproved the concept of spontaneous generation.

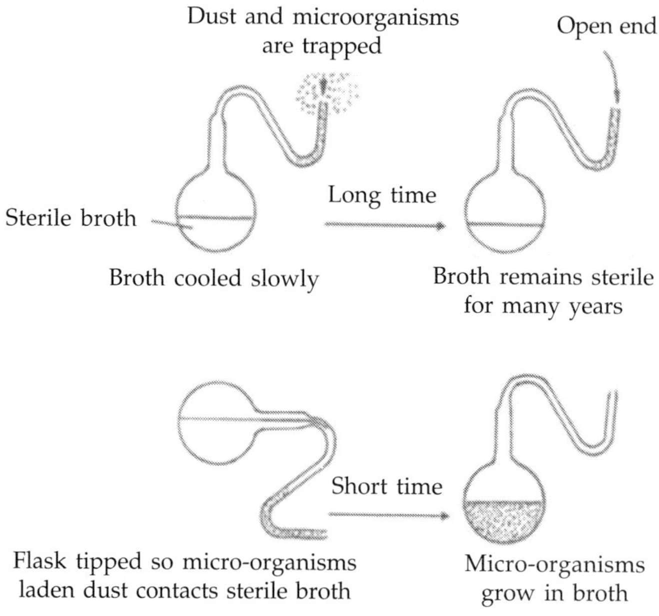

Louis pasteur's Experiment

The riddle was solved in 1876 when a German botanist, **Ferdinand Cohn** identified species of Bacillus, common to hay and cheese that contained heat-resistant endospores.

Theodor Schwann (1810-1882) passed air into flasks through red-hot tubes and no microbial growth was observed (Heat sterilization of air). He too, confirmed that air carries microbes which causes contamination and disproved the spontaneous generation theory.

III. Development of the Germ Theory of Disease

Joseph Lister, a British physician, in the mid 1800's, developed antiseptic surgery, which included heat-sterilization of instruments and application of phenols to wounds and dressings.

Robert Koch (1843 – 1910), provided a critical link between microbes and disease when he used a series of postulates to identify the cause of anthrax. Koch's postulates are still in use today in order to prove the cause of an infectious disease.

Koch's postulates - To prove that a disease is caused by an infectious agent

1. The agent must be present in every case of the disease.
2. The agent must be isolated and cultured *in vitro.*
3. The disease must be reproduced when a pure culture of the agent is inoculated into a susceptible host.
4. The agent must be recoverable from the experimentally-infected host.

Further Contributions of Koch

- Isolated the bacteria that cause cholera and tuberculosis.
- Developed tuberculin, now used in a skin test for TB (originally intended for use as a vaccine against TB).
- Developed acid-fast staining technique.
- Identified bacterial endospores.
- With colleagues, he was the first to grow cultures on solid media.
- Developed pour plate technique.
- Received noble prize for medicine in1905.

Only when all 4 of these steps have been fulfilled the original agent can be identified as the causative agent of the disease.

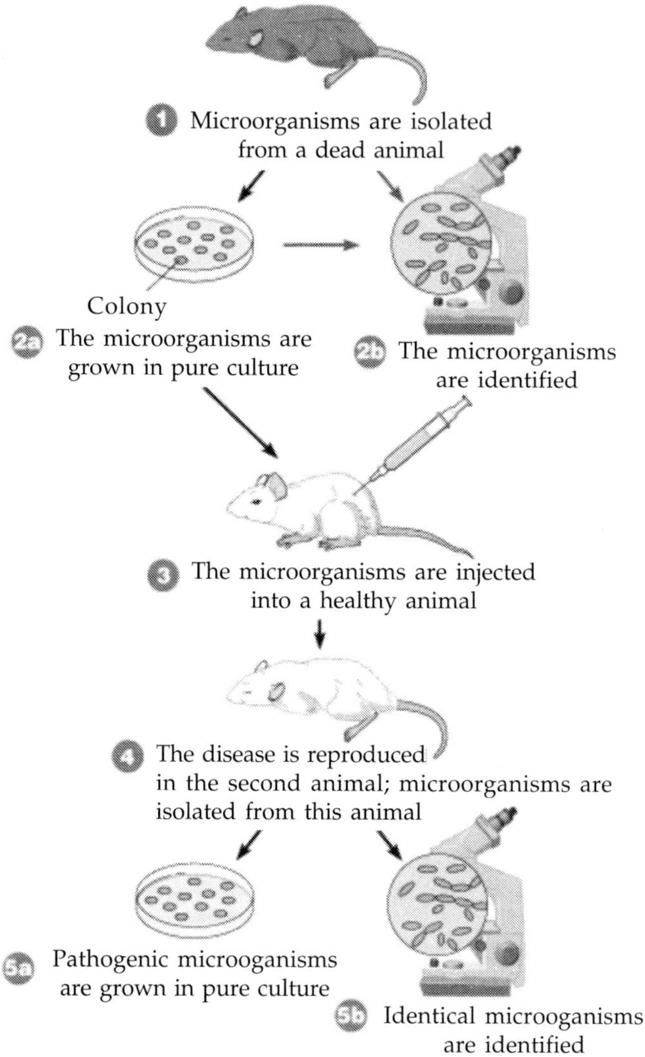

Kotch's postulate

The actual progress in the development of microbiology depended on

1. Invention and progressive evolution of microscope.
2. The use of stains and their development for the study of morphology.

3. Finding artificial media for studying cultural characteristics.
4. Development of science of immunology.
5. Introduction of chemotherapy - Discovery of Antibiotics.

Contributions of Louis Pasteur

- Disproved the Spontaneous generation theory.
- Discovered that fruits can be fremented to alcohol by microbes, which paved the way for fermentation.
- Sorted different microbes giving different taste of wine.
- He selected a particular strain (Yeast) for high quality wine.
- He developed a method to remove the undesired microbes from juice without affecting its quality. Heating the juice at 62.8°C for half-an hour did the job. This technique is called as Pasteurization, which is commonly used in the field of milk industry.
- He discovered that parasites (protozoa) causing pebrine disease of silk worm. He suggested that disease free caterpillars can eliminate the disease.

Pasteur

- He isolated the anthrax causing bacilli from the bloods of cattle, sheep and human being.
- He also demonstrated the virulence (ability of microbe to cause disease) of bacteria.
- He developed vaccine (a killed or attenuated microbe to induce the immunity) against rabbis from the brains and spinal cord of rabbit.

Further Inventions of Louis Pasteur

a) Fermentation

In 1857 Napoleon III requested help to search for the cause of wine spoilage. Pasteur was the first to discover specific biochemical reactions of microbes. He observed that yeast fermented sugar to ethanol and that bacteria oxidized alcohol to acetic acid to cause spoilage.

b) Pasteurization

Pasteur developed a heating process used to kill spoilage germs in wine but which preserved flavor. Prior to pasteurization diseases such as TB were also transmitted through milk. In the early 1900's Mycobacterial-contaminated milk was an important source of TB. Pasteurization also eliminates the milk-borne pathogen *Brucella*. Pasteur also made huge contributions to the field of vaccination and immunity.

c) Vaccination and Immunology

A vaccine is a preparation of microbes or their subunits that is designed to produce immunity to a disease. The ancient Chinese were among the earliest cultures to recognize that material from recovering smallpox victims could be used to immunize others. The problem with crude techniques such as these is that they used living organisms for immunizations, which could potentially be virulent. Louis Pasteur also made large contributions with respect to vaccines. Pasteur found that cholera organisms lost their virulence when passaged in culture. Attenuated (weakened) organisms inoculated into chickens could protect them from virulent strains. Pasteur also developed early vaccines against rabies and anthrax. Ironically, during the course of his work, he lost his father and two sisters to typhoid fever.

Pasteur's demonstration on immunization

During 1880, he isolated the bacterium causing chicken cholera. He arranged for a public demonstration and inoculated healthy chicks with his pure cultures. The chickens failed to get sick and died. He found that the cultures he used was several weeks old. Some weeks later he repeated the experiments. He used two groups of chickens and inoculated with fresh cultures. The chicken in second groups got sick and died. The first group remained hale and healthy. He found that the bacteria could lose their ability to produce disease i.e., virulence. But this attenuated bacteria still retained their capacity for stimulating the host to produce substances i.e., antibodies, that protect against subsequent exposure to virulent organisms.

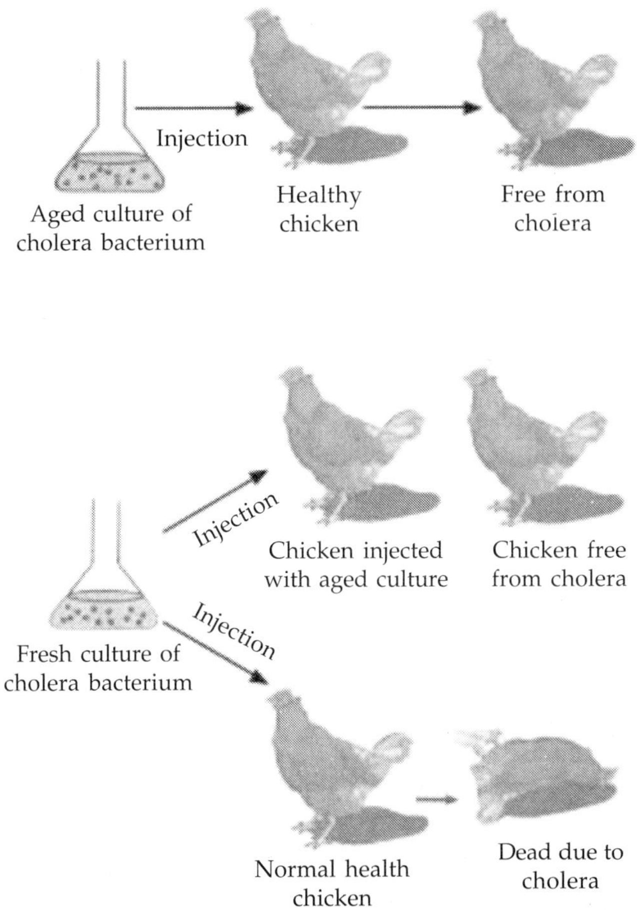

The classic experiment of Pasteur with chicken (fowl) cholera

Development of rabies vaccine

- Rabies (hydrophobia), a disease transmitted to people by bites of dogs, Cates and other animals
- Pasteur produced this disease in rabbits by inoculating the saliva from mad dogs
- He removed the brain and spinal cord from the infected rabbits, dried, pulvarised and mixed with glycerin
- Injecting this mixture into dogs protected them against rabies
- Joseph Meister, A boy bitten by a mad wolf was injected with same mix and he was saved after several weeks

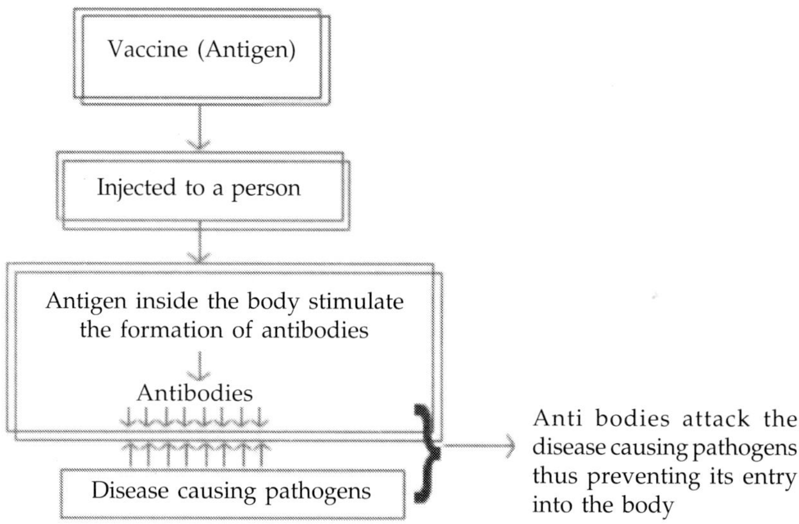

Other contributions from renowned scientists

Edward Jenner (1749-1823) demonstrated that inoculation with scrapings from cowpox provided immunity to the more virulent smallpox virus. He used an 8 year old boy called James Phipps to test his vaccine. James was inoculated with virulent matter from cowpox lesions on the fingers of a milkmaid called Sarah Neimes. Phipps developed a mild fever and some cowpox lesions. James was protected from the disease. Jenner noted that this immunity was "a change, which endures throughout life." Jenner developed vaccines for small pox disease. He coined the term vaccines (vacca meaning 'cow') and is considered as "**Father of Immunology**".

Elie Metchnikoff was a Russian researcher and is considered the father of cellular immunity largely for his observations of phagocytosis in sea star larvae. Metchnikoff shared the Nobel Prize for medicine with the German microbiologist, Paul Ehrlich in 1908.

Joseph Lister (1878), developed **Pure culture** technique. Pure culture referred as the growth of mass of cells of same species in a vessel. He developed the pure cultures of bacteria using serial dilution technique. By spraying carbolic acid on surgical instruments, wounds and dressings, he reduced surgical mortality due to bacterial infection considerably. He is considered as the **"Father of antiseptic surgery".**

Walther Hesse & his wife Fannie E. Hesse (1883), used agar instead of gelatin for preparation of culture media. Agar goes to solution at 100°C and solidifies at 45°C. Till now agar is not replaced by any other solidifying substance.

Alexander Fleming (1928), discovered antibiotic penicillin, which is an important milestone in medical microbiology. He found that natural substances having antimicrobial activity is found in saliva and nasal mucous. He also identified *Staphylococcus aureus.*

Florey & Chain - isolated Penicillin in pure culture

John Tyndall, devised a method of killing the spores that came to be known as "Tyndallization". He showed that carbon dioxide and ozone absorb heat radiation much more strongly than the rest of the atmosphere and thus formulated the basis for Green house effect

Selman A Waksman, coined the term antibiotics (referring a chemical substance of microbial origin which is in small quantity exert antimicrobial activity). During 1927, he wrote the book on Principles of soil Microbiology. In 1939 Waksman and his colleagues undertook a systematic effort to identify soil organisms producing soluble substances that might be useful in the control of infectious diseases, what are now known as antibiotics. Within a decade ten antibiotics were isolated and characterized by him, three of them with important clinical applications, Actinomycin in 1940, Streptomycin in 1944, and Neomycin in 1949. Eighteen antibiotics were discovered under his general direction.

Martinus W. Beijerinck, (1851-1931), developed the enrichment culture technique, simultaneously with Sergey Winogradsky, which permits the isolation of highly specialized microorganisms. Beijerinck cultivated and isolated *Bacillus radicicola (*later named as *Rhizobium leguminosarum),* a bacteria that fixes free nitrogen and causes the formation of nodules on the roots of Legumes. He also characterized *Azobacter* as nitrogen-fixing, and isolated the new genus, *Aerobacter*. Isolated sulphur reducing bacteria and sulphur oxidizing bacteria from soil. In studying tobacco mosaic disease, he concluded that the filterable pathogen was a *contagium vivum fluidum,* a term coined to convey his concept of a living infectious agent in a fluid (noncellular) form.

Sergey Winogradsky (1856-1953), developed Enrichment Culture Technique for the isolation of *Beggiatoa* sp. He explained the chemoautotrophic nature of bacteria and initially called them as "anorgoxydants." He also identified the process of Nitrification; isolated the nitrifying bacteria *Nitrobacter* and isolated *Azotobacter chroococcum* and proved its nitrogen fixing capacity. Due to his varied contributions in soil microbiology he is considered as the **"Father of Soil Microbiology".** Additionally, he identified the Chemolithotrophic nutrition of soil bacteria and also discovered anaerobic nitrogen fixing bacterium *Clostridium pasteurianum.*

S.N.Winogradsky

Alexander Fleming

S.A. Waksman

□□□

Chapter 3

Bacterial Morphology

Morphology is the branch of biology that deals with the study of the forms of living organisms with relationships between their structures. Major characteristics of bacterial cells, size, shape, structure and arrangement constitute the morphology of the cell. This chapter mainly throws light on all aspects of the bacterial morphology.

The size, shape and arrangement of bacterial cells

Size

Bacteria range in size from approximately as small as the largest viruses to large enough for single cells to be visible by the naked eye. The size is, from about 0.1 to about 600 µm over a single dimension. Bacteria vary in size as much as in shape. The smallest (e.g., some members of the genus *Mycoplasma*) are about 100 to 200 nm in diameter, approximately the size of the largest viruses (poxviruses). *Escherichia coli*, a bacillus of about average size, is 1.1 to 1.5 µm wide by 2.0 to 6.0 µm long. A few become fairly large; some spirochetes occasionally reach 500 µm in length, and the Cyanobacterium - *Oscillatoria* is about 7 µm in diameter (the same diameter as a red blood cell).

Recently a huge bacterium has been discovered in the intestine of the brown surgeonfish, *Acanthurus nigrofuscus*. *Epulopiscium fishelsoni* grows as large as 600 μm by 80 μm, a little smaller than a printed hyphen.

It is now clear that few bacteria are much larger than the average eukaryotic cell. *Epulopiscium fishelsoni* is an example of a very big prokaryote. *Epulopiscium fishelsoni* (is) a prokaryote related to the gram-positive genus *Clostridium*. In stark contrast to the large size of *Epulopiscium fishelsoni*, there exists an extremely small (i.e., bacteria size) eukaryote, *Nanochlorum eukaryotum*. '*Nanochlorum eukaryotum*' is only about 1 to 2 μm in diameter (about the size of *E. coli*), yet is truly eukaryotic and has a nucleus, a chloroplast, and a mitochondrion.

Surface to volume ratio

Because of their small size, bacteria have a large surface-to-volume ratio.

Ratio = surface area/volume

In bacteria, surface area is more and volume is less and the ratio is greater than 1 (positive) in bacteria. For example: Spherical bacteria with a diameter of 2 μm have a surface area of about 12 μm^2 and a volume of about 4 μm^3. Their surface-to-volume ratio is 12:4 or 3:1.

The large surface-to volume ratio of bacteria means that no internal part of the cell is very far from the surface and that nutrients can easily and quickly reach all parts of the cell. Large surface area bacteria cell has easy respiration and discharge the waste materials very quickly. The large *surface to volume ratio* seen in bacteria is one reason that prokaryotes are so successful despite their relatively simple morphologies.

Shape

The shape of a bacterium is governed by its rigid cell wall however exactly what attribute of this rigid material determines that a cell will have a particular shape is not yet understood. Typical bacterial cells are spherical (**cocci**: singular. coccus); straight rods (**bacilli**; singular. bacillus); or rods that are helically curved (**spirilla**: singular. spirillum). Although most bacterial species have cells that are of a fairly constant and characteristic shape, some have cells that are **pleomorphic**, i.e., that can exhibit a variety of shapes.

Arrangement

Bacterial cells are usually arranged in a manner characteristic of their particular species. Although it is rare that all the cells of a species are arranged in the same manner, it is the predominant arrangement that is the important feature.

1. *Cocci* appear in several characteristic arrangements, depending on the plane of cellular division and whether the daughter cells stay together following division as follows:

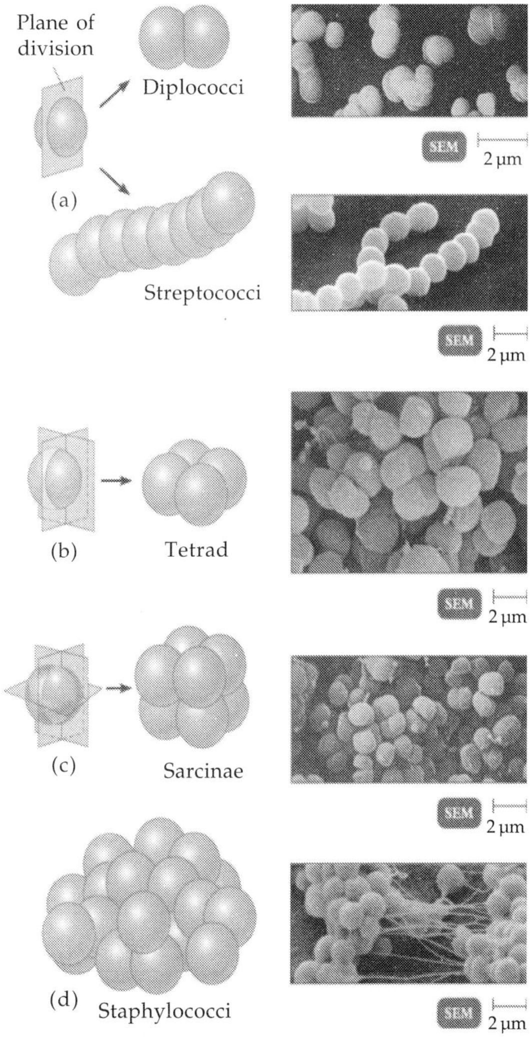

   (a) *Diplococci* - Cells divide in one plane and remain attached in pairs. e.g., *Neisseria gonorrhoeae.*

   Streptococcus - Cells divide in one plane and remain in chains of 4-20. e.g., *Lactococcus spp, Enterococcus spp.*

   (b) *Tetrads or Tetracocci* - Cells divide in two planes and remains as four cells. e.g., *Micrococcus luteus.*

   (c) *Sarcinae* - Cells divide in three planes in a regular pattern producing cuboidal arrangement of cell. e.g., *Sarcina.*

   (d) *Staphylococci* - Cells divide in three planes in an irregular pattern producing bunches of cells.

2. *Bacilli* are longer than they are wide and lack extreme curvature variations on rod shaped bacteria: are rod, tapered rod, staff, cigar, oval, curved. Bacilli typically divide only across their short axis. The arrangements of bacilli are as follows:

   (a) Bacilli - single rod

   (b) *Diplobacilli* - Two bacilli together

   (c) *Streptobacilli* - Chains of bacilli. e.g., *Bacillus megaterium*

   (d) *Coccobacillus* are not perfectly round, as are true cocci, but appear to have blunted ends

   (e) *Trichomes* - Similar to chains but have a much larger area of contact between the adjacent cells. e.g., *Beggiatoa* and *Saprospira* species

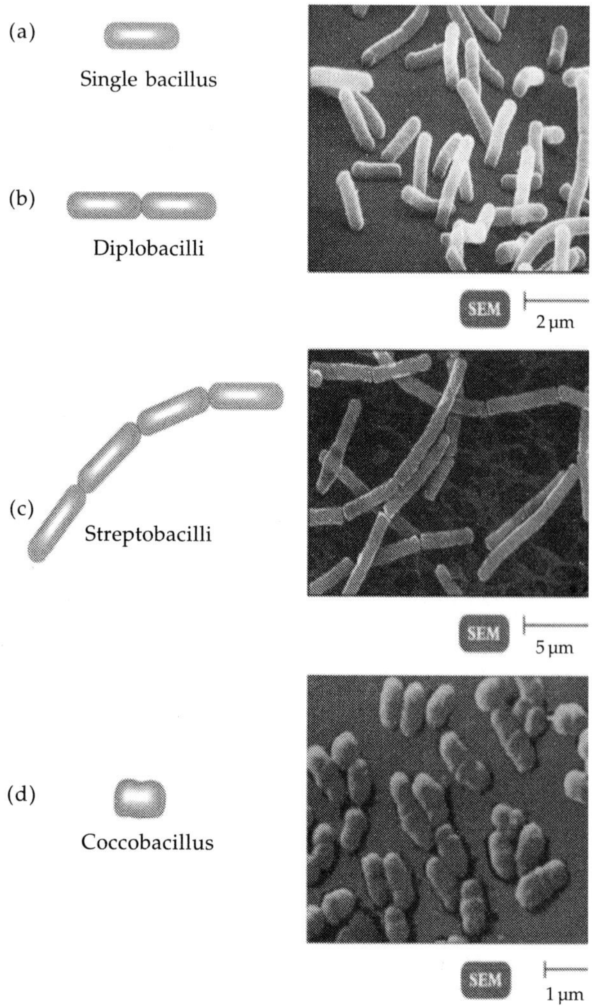

   (f) *Palisades* - Cells are lined side by side like matchsticks at angles to one another. e.g., *Corynebacterium diphtheria*

Still others, such as *Streptomyces* species, form long, branched, multinucleate filaments called hyphae (singular, hyphae) which collectively form a mycelium.

3. Spirilli are curved bacteria and vary with the number of twists they have. Depending upon the twist, they are of different types as given below:
   (a) *Vibrioid* - Spirals with less than one complete twist or comma shaped. e.g., *Vibrio*.
   (b) *Spirillium* - Spiral shape that is rigid (i.e. rigid helix)
   (c) *Spirochete* - If the organism is flexible and undulating (i.e. flexible helix)

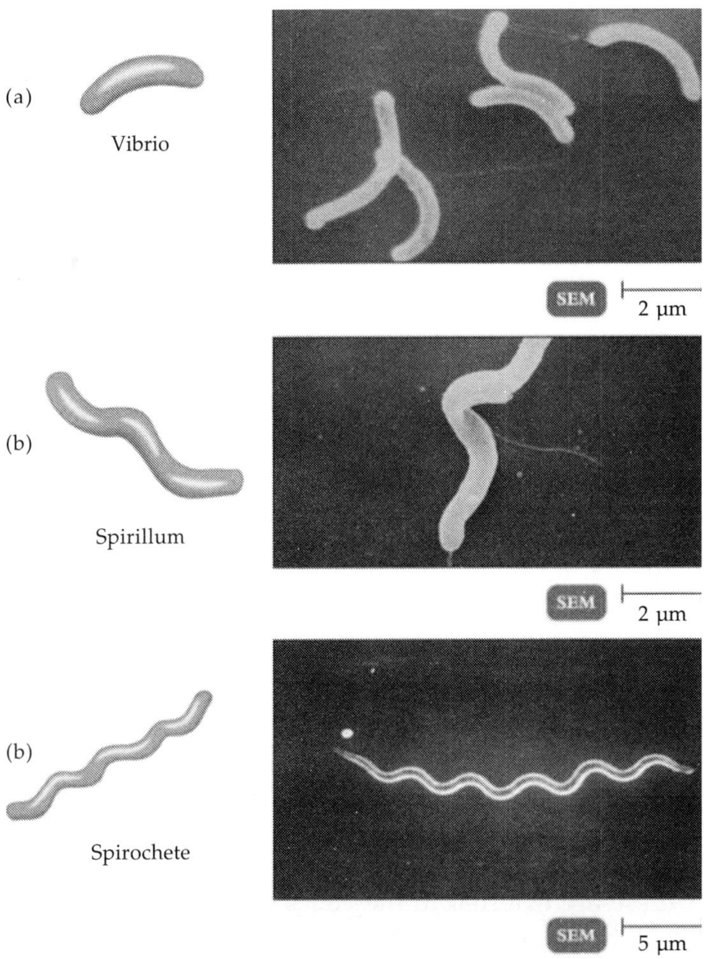

Other shapes

In addition to the common bacterial shapes, many others also occur: as pear shaped cells (e.g., *Pasteuria);* lobed spheres (e.g., *Sulfolobus);* rods with square rather than the usual hemispherical ends (e.g., *Bacillus anthracis*); stacks of coins (e.g., *Caryophanon*); rods with helically sculpture surfaces (e,g., *Seliberia);* and many others like square shape in Archaebacteria.

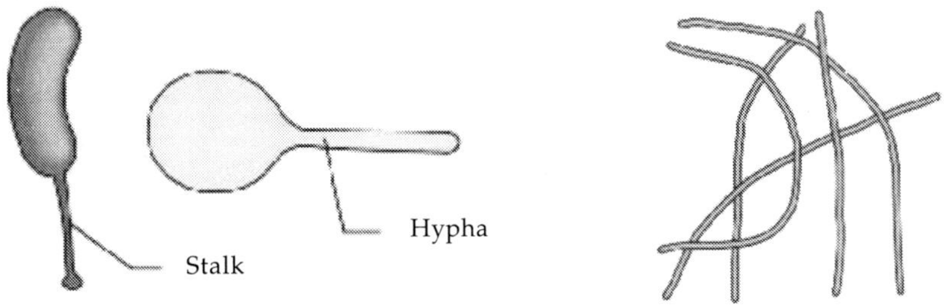

Budding and appendaged bacteria Filamentous

Filamentous bacteria

In addition to the ususal morphological shapes, several groups of bacteria are immediately recognized by their unusual shapes. Examples include, **appendage bacteria,** which possess extensions of their cells as long tubes or stalks and **filamentous bacteria,** which form long, thin cells or chains of cells.

Composition and function of bacterial structures

Bacteria are very small. Yet despite their size they show a surprising degree of complexity in their structures. Knowing the structure of a microbe helps in understanding how a microbe functions.

Structurally, a typical bacterium usually consists of:

a. A cytoplasmic membrane surrounded by a peptidoglycan cell wall and maybe an outer membrane;

b. A fluid cytoplasm containing a nuclear region (nucleoid) and numerous ribosomes; and

c. Often various external structures like a glycocalyx, flagella, and pili.

Examination of bacterial cell reveals various component structures. Some of these are external to the cell wall and others are internal to the cell wall. Some structures are present in only certain species.

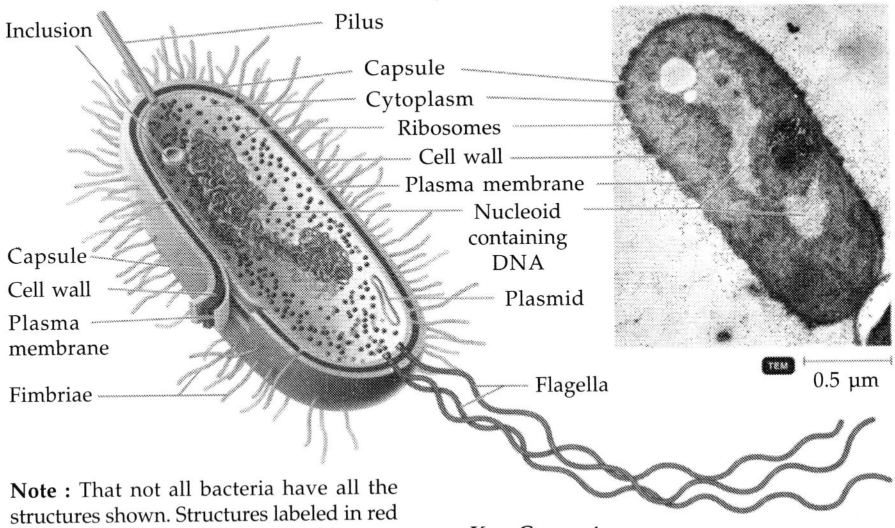

Note : That not all bacteria have all the structures shown. Structures labeled in red are found in all bacteria. Both the drawing and the micrograph show a bacterium sectioned lengthwise to reveal the internal composition.

Key Concept
Prokaryotic cells lack membrane-enclosed organelles. All bacteria contain cytoplasm, ribosomes, a plasma membrane, and a nucleoid. Almost all bacteria have cell walls.

I. Structures external to the cell wall

1. Flagella
2. Fimbriae and Pili
4. Capsule
5. Sheath
6. Prostheca

1. Flagella

Bacterial flagella (singular. flagellum) are hairlike, helical appendages that protrude through the cell wall and are responsible for swimming motility. They are much thinner than the flagella or cilia of eukaryotes, being 0.01 to 0.02 μm in diameter, and they are also much simpler in structure. Their location on the cell varies depending on the bacterial species and may be polar at one or both ends of the bacterium) or lateral (along the sides of the bacterium).

Flagella structure

It consists of 3 main parts

1. Basal body
2. Hook
3. Filament

Flagellar Structure

Flagella are not straight but helically shaped; when flattened, they show a constant distance between two adjacent curves, called the *wavelength,* and this wavelength is constant for a given organism. The filament of bacterial flagella is composed of subunits of a protein called *flagellin.* The shape and wavelength of the flagellum are in part determined by the structure of the flagellin protein and also to some extent by the direction of rotation of the filament. The basic flagellar structure to be described here varies little among species of *Bacteria,* however, in *Archaea* several different flagellins are known, and the flagellar structure is probably quite different from that of *Bacteria.* In species of the latter, flagellin is highly conserved, suggesting that flagellar motility has deep evolutionary roots within this prokaryotic group.

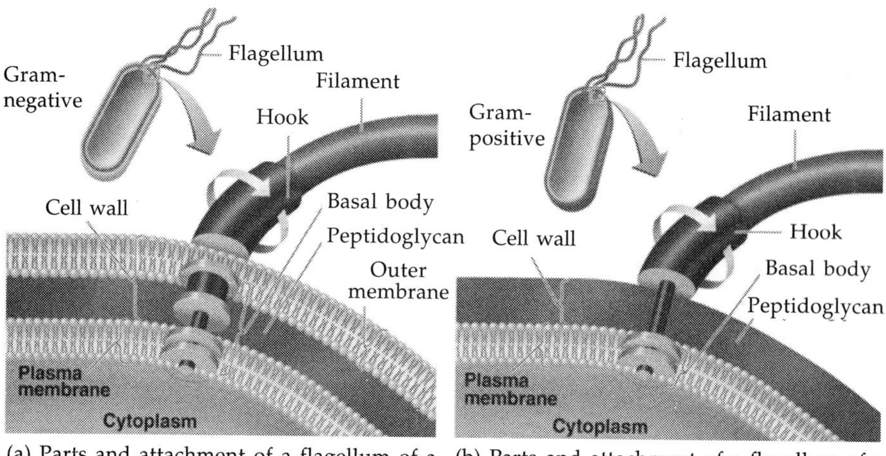

(a) Parts and attachment of a flagellum of a gram-negative bacterium (b) Parts and attachment of a flagellum of a gram-positive bacterium

The base of the flagellum is different in structure from that of the filament. There is a wider region at the base of the flagellum called the *hook.* The hook consists of a single type of protein and functions to connect the filament to the motor portion of the flagellum. The motor is anchored in the cytoplasmic membrane and cell wall. The motor consists of a small central rod that passes through a system of rings. In gram-negative *Bacteria,* an outer ring is anchored in the lipopolysaccharide layer and another in the peptidoglycan layer of the cell wall, and an inner ring is located within the cytoplasmic membrane. In gram-positive *Bacteria,* which lack the outer lipopolysaccharide layer, only the inner pair of rings is present.

Surrounding the inner ring and anchored in the cytoplasmic membrane are a pair of proteins called Mot. These proteins actually drive the motor causing a torque that rotates the filament. A final set of proteins, called the Fli proteins function as the motor switch, reversing rotation of the flagella in response to intracellular signals.

Unlike a hair, a flagellum grows at its tip rather than at the base. Flagellin monomers synthesized within the cell are believed to pass along the hollow center of the flagellum and are added to the distal end of the filament.

Flagella Movement

Other types of movement in bacteria

i. Gliding motility
ii. Chemotacitc movement
iii. Magnetotactic movement
iv. Phototactic behaviour

Flagellar arrangement in bacteria

1. Monotrichous - Single polor e.g., *Pseudomonas aeruginosa*
2. Lophotrichous - Cluster of polar flagella at one end. e.g., *Pseudomonas fluorescens*

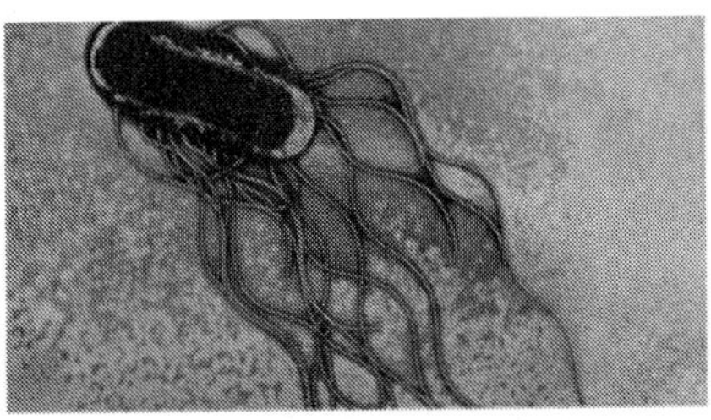

(a) Peritrichous SEM 0.5 µm

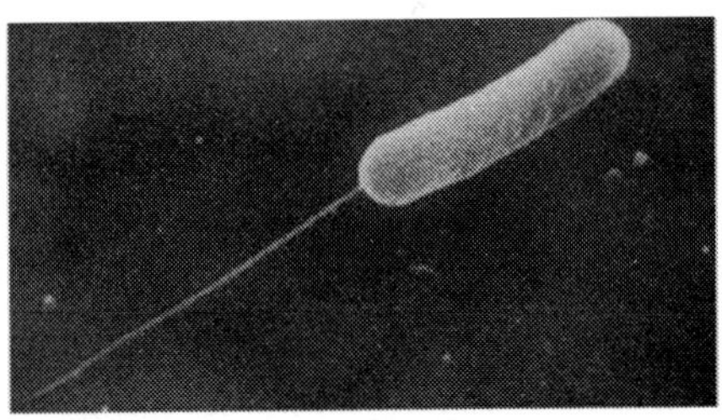

(b) Monotrichous and polar SEM 0.5 µm

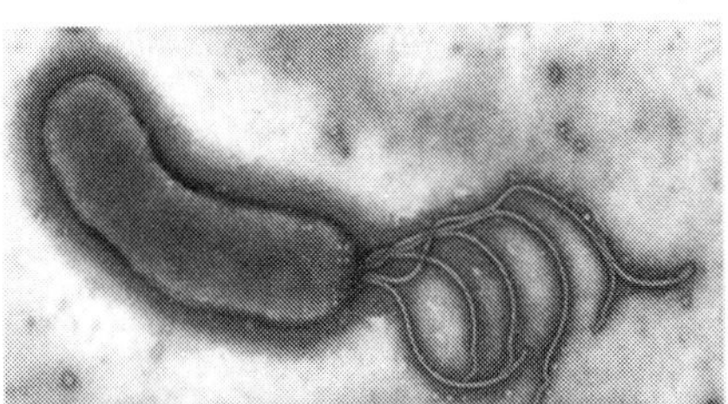

(c) Lophotrichous and polar SEM 0.5 µm

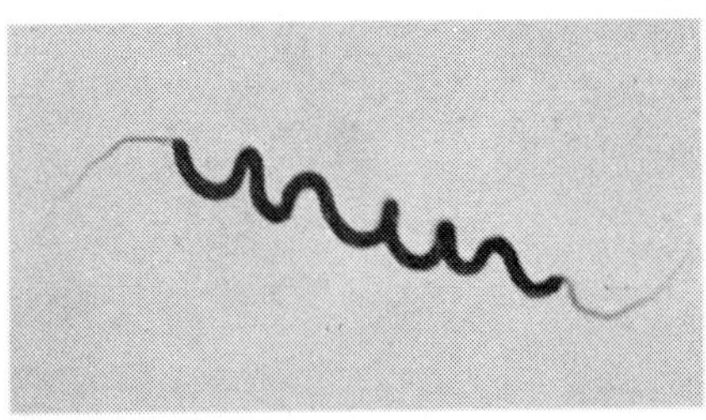

(d) Amphitrichous and polar SEM 5 µm

3.	Amphitrichous -	Single or cluster at both poles	e.g., *Aquaspirillum serpens*
4.	Peritrichous -	Surrounded by lateral flagella	e.g., *Salmonella typhi*

The flagella are responsible for motility of the bacterium. It rotates like a propeller. Rings in the basal body rotate relative to each other causing the flagella to turn. The energy to drive the basal body is obtained from the proton motive force. The average speed of bacterial movement is 50 mm/sec which is equivalent to its 10 times body length.

Endoflagella

The flagella or filaments, present between the outer membrane and inner membrane of the cell wall of the bacterium are called as *endoflagella* or *periplasmic flagella* or *axial filaments*. They are attached at one end of the cell. e.g., Spirochetes - group of bacteria; Bacterium - *Spirochete, Leptospira.*

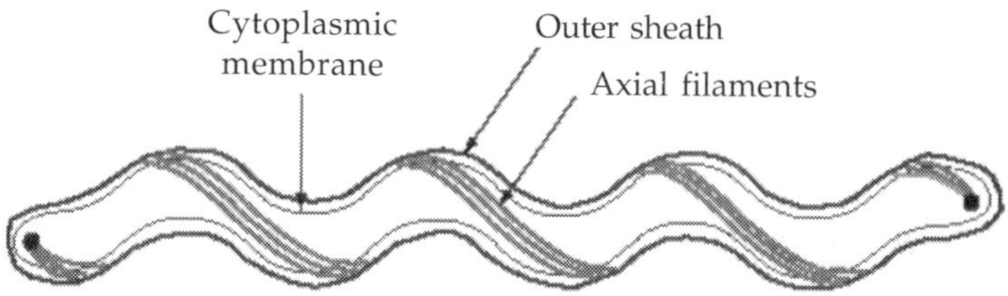

2. Fimbriae and Pili

Fimbriae and *pili* are structurally similar to flagella but are not involved in motility. Fimbriae are considerably shorter than flagella and are more numerous but, like flagella, made of protein. The

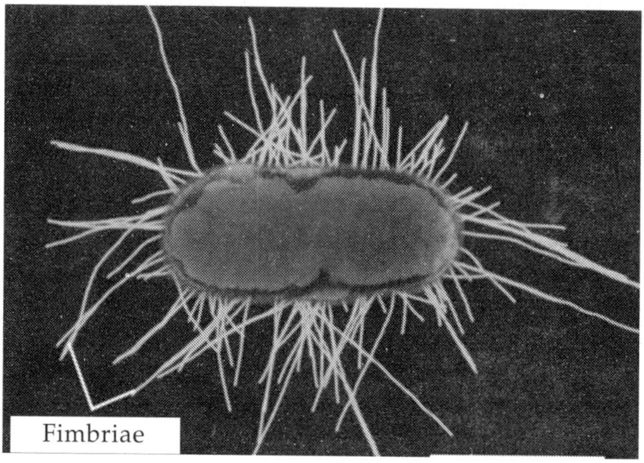

functions of fimbriae are not known for certain, but there is good evidence that they enable organisms to stick to surfaces, including animal tissues in the case of some pathogenic bacteria, or to form pellicles or biofilms on surfaces.

Pili (singular, pilus) are hollow, nonhelical, filamentous appendages that are thinner, shorter, and more numerous than flagella. They do not function in motility, since they are found on nonmotile as well as motile species. There are, however, several functions associated with different types of pili. One type, known as the F pilus (or sex pilus), 'serves as the port of entry of genetic material during bacterial mating. Some pili play a major role in human infection by allowing pathogenic bacteria to attach to epithelial cells lining the respiratory, intestinal, or genitourinary tracts. This attachment prevents the bacteria from being washed away by the flow of mucous or body fluids and permits the infection to be established.

3. Capsules

Some bacterial cells are surrounded by a viscous substance forming a covering layer or envelope around the cell wall. If this layer can be visualized by light microscopy using special staining methods, it is termed as capsule. If the layer is too thin to be seen by light microscopy it is termed a microcapsule; if it is so abundant that many cells are embedded in a common matrix, the material is called slime.

By light microscopy, capsules appear to be amorphous gelatinous areas surrounding a cell however, special techniques designed to preserve delicate structures for observation by electron microscopy have revealed that capsules consist of a mesh or net work of line strands.

In many instances capsular material is not highly water-soluble and therefore does not readily diffuse away from the cells that produce it. In other instances the material is highly water-soluble and dissolves in the medium, sometimes dramatically increasing the viscosity of the broth in which the organisms are cultured.

Capsules can serve a number of functions, depending on the bacterial species. (1) They may provide protection against temporary drying by binding water molecules. (2) They may block attachment of bacteriophages. (3) They may be anti phagocytic; i.e., they inhibit the engulfment of pathogenic bacteria by white blood cells and thus contribute to invasive or infective ability (virulence). (4) They may promote attachment of bacteria to surfaces; for example, *Streptococcus* mutans, a bacterium associated with 'producing dental caries firmly adheres to the smooth surfaces of teeth because of its secretion of a water-insoluble capsular glucan. (5) If capsules are composed of compounds having an electrical charge, such as sugar-uronic acids, they may promote the stability of bacterial suspension by preventing the cells from aggregating and settling out, because cells bearing similarly charged surfaces tend to repel one another.

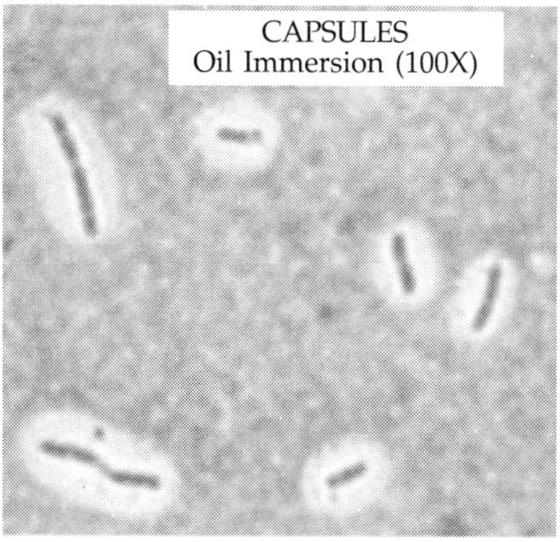

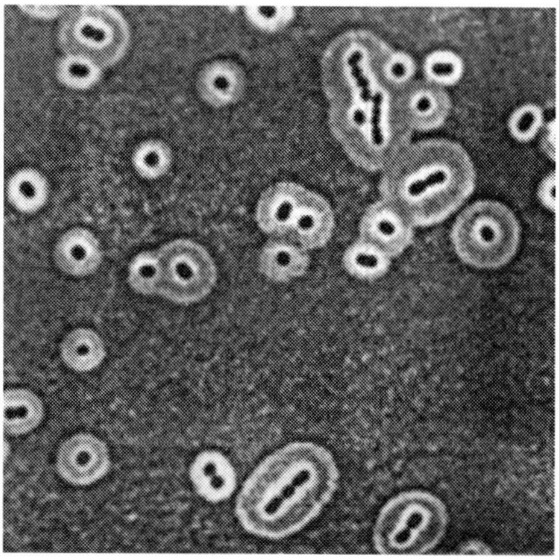

Most bacterial capsules are composed of polysaccharides. Capsules composed of a single kind of sugar are termed homopolysaccharides; are usually synthesized outside the cell from disaccharides by exocellular enzymes. The synthesis of glucan (a polymer of glucose) from sucrose by *Strptococcus mutans* is an example. Other capsules are composed of several kinds of sugars and are termed hetero-polysaccharides; these are usually synthesized from sugar precursors that are activated (energized) within the cell attached to a lipid carrier molecule, transported across the cytoplasmic membrane, and polymerized outside the cell. The capsule of *Klebsiella pneumoniae* is an example.

A few capsules are polypeptides. For example, the capsule of the anthrax organism, *Bacillus anthracis,* is composed entirely of a polymer of glutamic acid. Moreover, this peptide is an unusual one because the glutamic acid is the rare D optical isomer rather than the usual L isomer commonly found in nature.

4. Sheath

Some species of bacteria; particularly those from freshwater and marine environments, form chains or trichomes that are enclosed by a hollow tube called a sheath. This structure is most readily visualized when some of the cells have migrated from it. Sheaths may sometimes become impregnated with ferric or manganese hydroxides, which strengthen them.

5. Prosthecae and stalks

Prosthecae are semi rigid extensions of the cell wall and cytoplasmic membrane and have a diameter that is always less than that of the cell. They are characteristic of a number of aerobic bacteria from freshwater and marine environments. Some bacterial genera such as *Caulobacter* have a single prostheca; others such as *Shigella* and *Ancalomicrobium* have several. Prosthecae increase the surface area of the cells for nutrient absorption which is advantageous in dilute environments. Some prosthecate bacteria may form a new cell (bud) at the end of a prostheca; others have an adhesive substance at the end of a prostheca that aids in attachment to surfaces.

Although the term stalk is sometimes used interchangeably with the terms prostheca or hypha. It is perhaps better to restrict its use to certain nonliving ribbonlike or tubular appendages that are excreted by the cell. Those stalks found in the genera *Gallionella* or *Planctomyces,* aid in attachment of the cells to surfaces.

II. Cell wall of bacteria

All the bacteria have rigid cell wall. The cell wall is the essential structure that protects the cell from mechanical damage and from osmotic rupture of lysis. Prokaryotes live in relatively diluted atmosphere (has lower osmotic pressure than the osmotic pressure of inside the cell), which accumulate high salt concentration inside. The osmotic pressure against the inside of the plasma membrane may be the equivalent to 10-25 atm. Since the plasma membrane is delicate, plastic structure, it must be restrained by an outside wall made of porous, rigid material that has high tensile strength. The cell wall of bacteria deserves special attention for several reasons.

- They are essential for viability
- They are composed of unique compounds nowhere else in nature
- They are one of the important site for attack by antibiotics
- They provide adherence and receptor sites for drug or virus
- They cause symptoms of disease in animals

- Protects the cells from bursting when pressure increases
- They provide immunological distinction and variation among strains of bacteria.

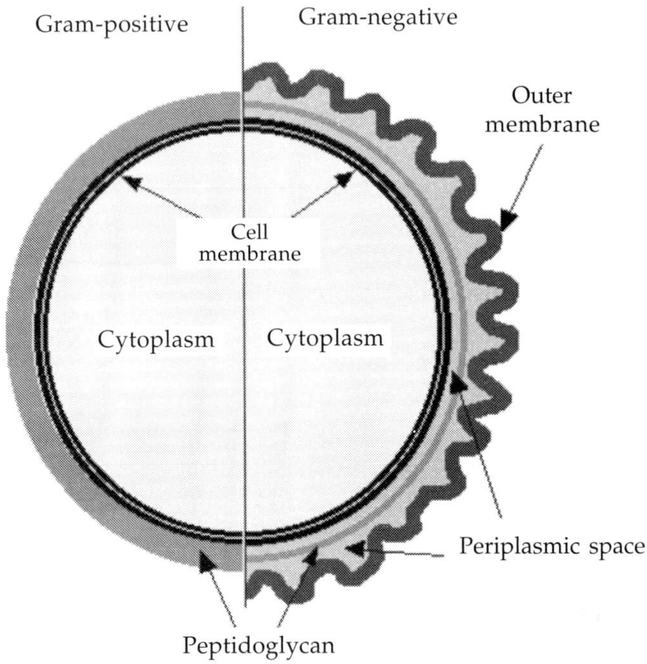

Cell wall difference between gram positive and gram negative bacteria

The bacteria can be divided in to two major groups based on their cell wall chemistry, called as gram positive and gram negative bacteria. The gram negative cell wall is a multilayered and quite complex, where as the gram positive cell wall consists of a single type of molecule and often much thicker.

a. Peptidoglycan layer

The rigid layer of both gram positive and gram negative bacteria is similar in chemical composition, called as peptidoglycan (or murein). Peptidoglycan is a thin sheet composed of 2 sugar derivatives, N-acetyl glucosamine and N-acetyl muramic acid and a group of amino acid side chain consisting of L-alanine, D-alanine, D-glutamic acid and either lysine or diaminopimelic acid (DAP). These constituents are connected to form a repeating structure, the glycan tetrapeptide (meso diaminopimelic acid, L-lysine, L-ornithine or L-diaminobutyric acid).

The basic structure of peptidoglycan is a thin sheet in which the glycan chains formed by the sugars are connected by peptide cross links formed by the amino acids. The glycosidic bonds connecting the sugars in the glycan chains are very strong, and also rigidity is more when these chains are cross linked by amino acids. In gram negative bacteria, cross linkage usually occur by direct peptide linkage of the amino group of DAP to the carboxyl group of the terminal D -alanine. In, gram positive bacteria, cross linkage is usually by a peptide interbridge, the kinds and numbers of cross linking amino acids vary from one organism to other. 90% of the cell wall consists of peptidoglycan and also teichoic acid is present in small amounts. Some bacteria have a single layer of peptidoglycan surrounding the cell, many bacteria especially gram positive bacteria have several (upto about 25) peptidoglycan layers. In Gram negative bacteria only about 10% of the wall is peptidoglycan, the majority of the wall consisting of a complex layer.

The peptidoglycan is signature of the bacteria can be destroyed by certain agents. Ex. Lysozyme, a protein that break the β-1, 4 glucoside linkage of peptidoglycan and weaken the cell wall. Water then enters to the cell, and cell swells and eventually bust. Lysozyme is present in the tears, saliva and body fluids of human.

If proper concentration of solute that does not allow the water to penetrate into the cell, such as sucrose is added to the medium, the solute concentration will be balanced between inside and outside the cell. Under those conditions, the lysozyme still digest the peptidoglycan, but lysis does not occur, instead, **protoplast** is formed. (Protoplast referred as the cell without cell wall, but osmatically protected). If these sucrose stabilized cells are placed in water, lysis occurs immediately.

Spheroplast often referred as the cell with some remaining of cell wall.

If peptidoglycan layer is removed in gram positive cell, protoplast will be formed and in gram negative bacteria will form spheroplast.

b. Outer membrane

Besides peptidoglycan layer, gram negative bacteria contain additional wall layer made of lipo-polysaccharide. The layer is called as lipopolysaccharide layer or LPS layer or outer membrane.

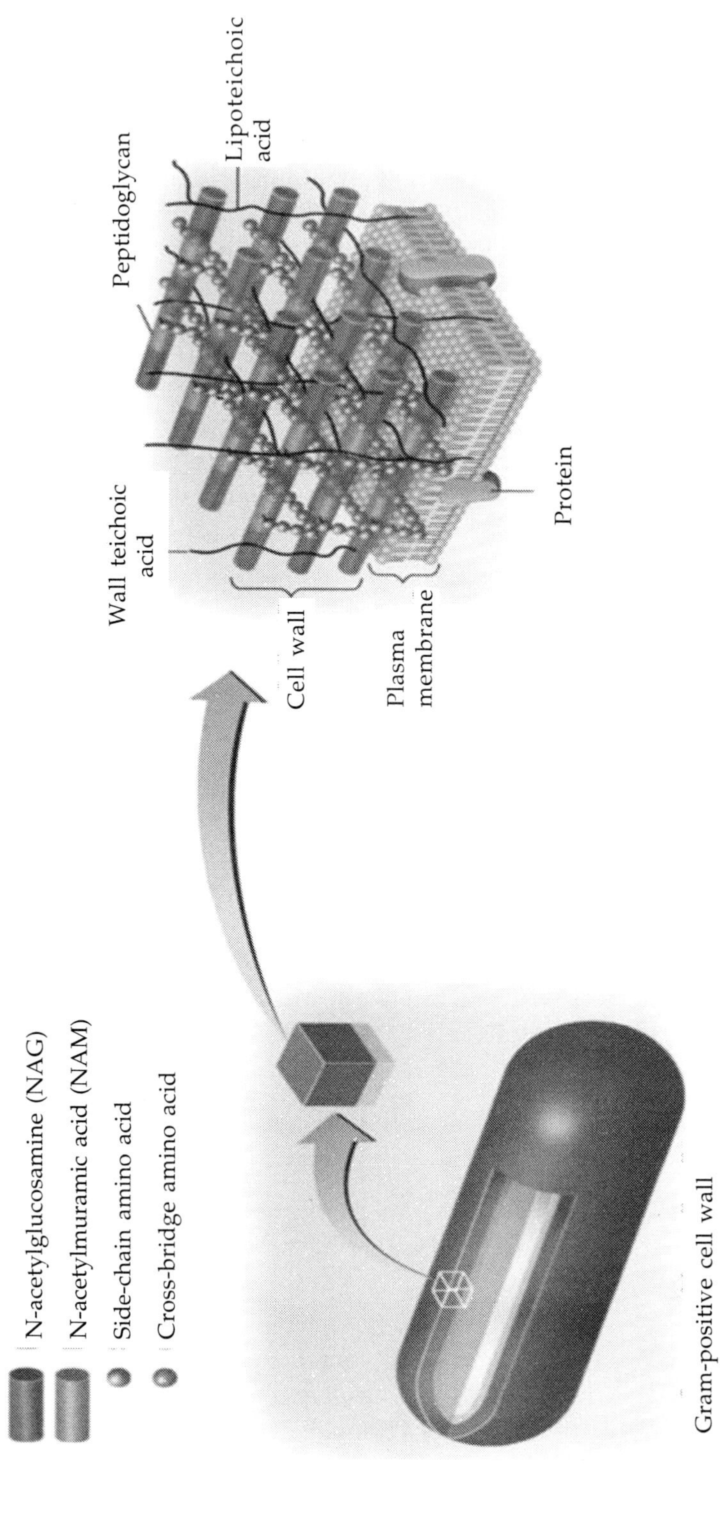
N-acetylglucosamine (NAG)
N-acetylmuramic acid (NAM)
Side-chain amino acid
Cross-bridge amino acid
Gram-positive cell wall
Wall teichoic acid
Peptidoglycan
Lipoteichoic acid
Cell wall
Plasma membrane
Protein

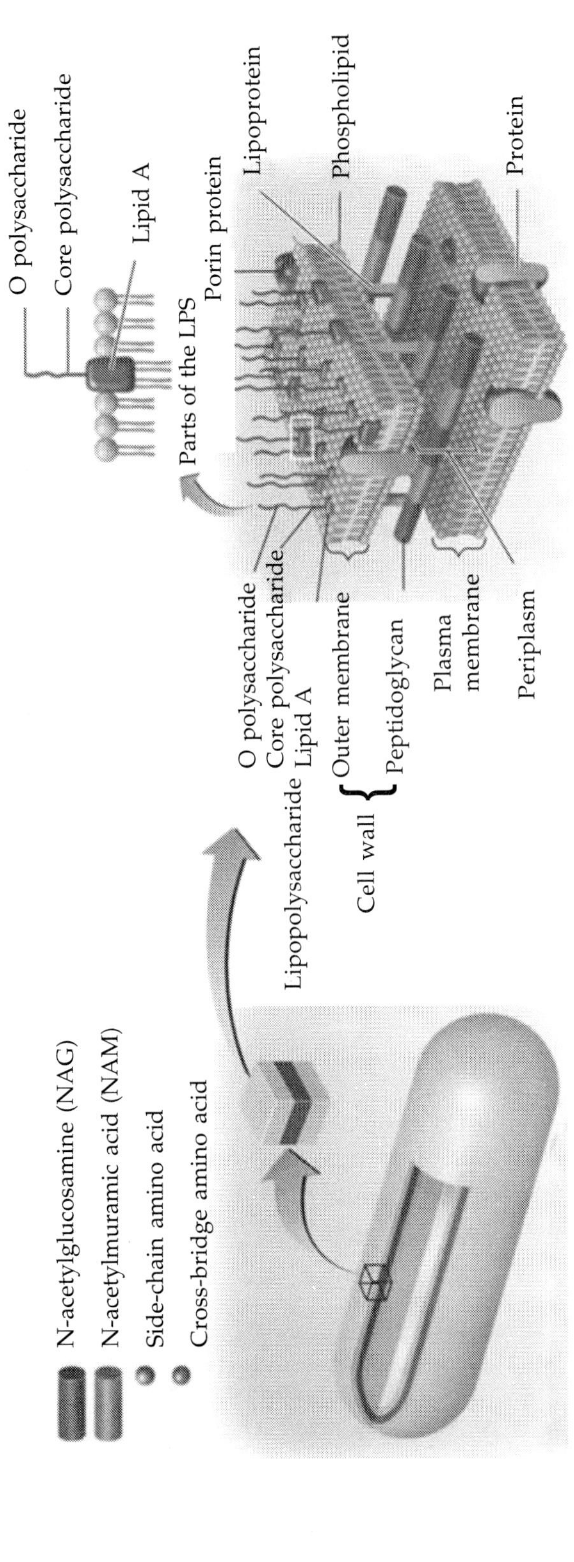

N-acetylglucosamine (NAG)
N-acetylmuramic acid (NAM)
Side-chain amino acid
Cross-bridge amino acid
Gram-positive cell wall
Lipopolysaccharide
O polysaccharide
Core polysaccharide
Lipid A
Cell wall
Outer membrane
Peptidoglycan
Plasma membrane
Periplasm
O polysaccharide
Core polysaccharide
Lipid A
Parts of the LPS
Porin protein
Lipoprotein
Phospholipid
Protein

The lipids and polysaccharides are intimately linked and formed this layer. The polysaccharide has two portions namely, o-polysaccharide and core polysaccharide.

O-polysaccharide contains galactose, glucose, rhamnose, mannose, and one or more dideoxy sugar like abequose, colitose, paratose or trylose. These sugars were connected four or five membered sequence which often branched. The core polysaccharide consists of keto deoxyoctonate, glucose, galactose and N-acetyl glucosamine.

The lipid portion of the LPS referred as lipid A, is not a glycerol lipid but instead, the fatty acids are connected by ester amine linkage to disaccharides composed of N-acetyl glucosamine. The fatty acids are caproic, lauric, mysteric, palmitic, stearic acids.

A lipid protein is also found on the inner side of the outer membrane of gram negative bacteria. It acts like an anchor between outer membrane and peptidoglycan layer. Most of the proteins embedded in the outer membrane are tenaciously held (integral proteins) or only loosely attached (peripheral proteins) and can be removed by mild treatments such as osmotic shock.

Endotoxins

In *Salmonella, Shigella, Escherichia,* the outer membrane causes toxic and pathogenic symptoms to human and mammals referred as endotoxins. This property is associated with lipid A.

c. Porins

Protein substance present in the outer membrane of gram negative bacteria. They act like channel for entrance and exit of hydrophobic low molecular substances.

d. Periplasm

The space in between peptidoglycan and outer membrane is referred as periplasm or periplasmic space. In gram positive bacteria, it refers the space between peptidoglycan and cell membrane. This space harbours hydrolytic enzymes, binding proteins, detoxifying enzymes etc.

III. Cell membrane

The cell membrane is the most dynamic structure in prokaryotes. They act as selective permeability barrier that regulates the passage of substance in and out of the cell.

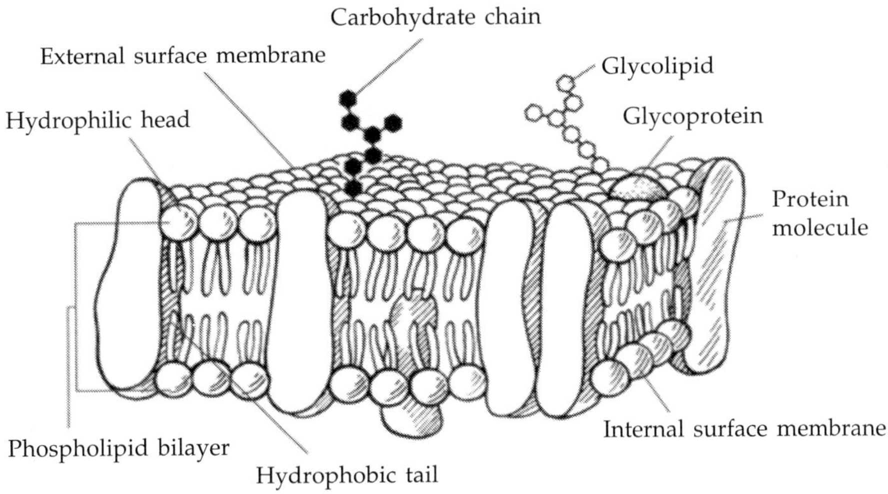

Fluid mosaic model of cytoplasmic membrane

Bacterial membrane consists of 40 per cent phospholipids and 60 per cent protein. The phospholipids forms bilayer and the proteins are present scatterly (called integral protein) and some protein present on the surface of the layer (called peripheral protein).

The phospholipids are phosphoglycerides in which straight chain fatty acids are ester linked to glycerol.

Fluid mosaic model of the cell plasma membrane

The plasma membrane that surrounds the cells has two layers (a *bilayer*) of *phospholipids* (fats with phosphorous attached. Each phospholipid molecule has a head that is attracted to water (*hydrophilic: hydro* = water; *philic* = loving) and a tail that repels water (*hydrophobic: hydro* = water; *phobic* = fearing). Both layers of the plasma membrane have the hydrophilic heads pointing toward the outside; the hydrophobic tails form the inside of the bilayer. Because cells reside in a watery solution (extracellular fluid), and they contain a watery solution inside of them (cytoplasm), the plasma membrane forms a circle around each cell so that the water-loving heads are in contact with the fluid, and the water-fearing tails are protected on the inside. Integral and pheripheral proteins become embedded in the bilayer, giving the membrane the look of a mosaic. Because the plasma membrane has the consistency of vegetable oil at body temperature, the proteins and other substances are able to move across it. That's why the plasma membrane is described using the *fluid*-mosaic model.

Functions of cell membrane

- Osmotic or permeability barrier
- Location of transport systems for specific solutes (nutrients and ions)
- Energy generating functions, involving respiratory and photosynthetic electron transport systems, establishment of proton motive force, and transmembranous ATP-synthesizing ATPase
- Synthesis of membrane lipids (including lipopolysaccharide in gram negative cells)
- Synthesis of murein (cell wall peptidoglycan)
- Assembly and secretion of extra cytoplasmic proteins
- Site for respiration
- Coordination of DNA replication and segregation with septum formation and cell division
- Chemotaxis (both motility and sensing functions)

IV. Structures internal to the cell wall

Apart from cell membrane, the following structures are seen.

1. Membranous intrusions
2. Cytoplasm
3. Cytoplasmic inclusions and vacuoles
4. Chromosomal DNA
5. Plasmid

1. Membraneous intrusions and intracellular membrane system

Bacterial cells do not contain membrane-enclosed organelles corresponding to the mitochondria and chloroplasts of eukaryotic cells. However, bacteria may have specialized, invaginations of the cytoplasmic membrane that can increase their surface area for certain functions.

Mesosomes

Many bacteria, especially Gram positive bacteria possess membrane invaginations in the form of systems of convoluted tubules and vesicles termed mesosomes. Those known as central mesosomes penetrate deeply into the cytoplasm, are located near the middle of the cell and seem to be attached to the cell's nuclear material; they are thought to be involved in DNA replication and cell division. In contrast, peripheral mesosomes show only a—shallow penetration into the cytoplasm and are not either restricted to a central location or associated with nuclear material; they seem to be involved in export of exocellular enzymes such as penicillinase.

Extensive intracellular membrane systems occur in methane-oxidizing bacteria, in certain chemoautotrophic bacteria and in nearly all phototrophic bacteria. They serve to increase surface area for various metabolic activities. For example, in phototrophic bacteria they are the site of the photosynthetic apparatus of the cell; the infoldings provide a large surface area to accommodate a high content of light-absorbing pigments. In the phototrophs known as cyanobacteria, special intracellular membranes (thylakoids) occur that seem to be separate from the cytoplasmic membrane. Also mesosomes are playing a major role in bacterial cell division and pigment production.

2. Ribosomes

The cell material bounded by the cytoplasmic membrane may be divided into (1) the **cytoplasmic area**, granular in appearance and rich in the macromolecular RNA-protein bodies known as ribosomes, on which proteins are synthesized; (2) the **chromatinic area**, rich in DNA; and (3) the fluid portion with dissolved substances. Unlike animal or plant cells, there is no endoplasmic reticulum to which ribosomes are bound; some ribosomes are free in the cytoplasm, and others, especially those involved in the synthesis of proteins to be transported out of the cell, are associated with the inner surface of the cytoplasmic membrane. When the ribosomes of procaryotes undergo sedimentation in a centrifuge, they have a sedimentation coefficient of 70 Svedberg units (70 S) and are composed of two subunits a 50 S and a 30 S subunit. This is in contrast to the ribosomes of eucaryotic organisms, which have a sedimentation coefficient of 80 S and are composed of a 60 S and a 40 S subunit.

3. Cytoplasmic inclusions and vacuoles

Concentrated deposits of certain substances are detectable in the cytoplasm of some bacteria. *Volutin* granules, also known as metachromatic granules, are composed of polyphosphate. They stain an intense reddish-purple color with dilute methylene blue and can be observed by light microscopy. By electron microscopy they appear -as round, dark areas. Volutin serves as a reserve source of phosphate.

(a) Small subunit (b) Large subunit (c) Complete 70S ribosome

Some inclusions in bacterial cells

Cytoplasmic inclusions	Where found	Composition	Function
glycogen	many bacteria e.g. *E. coli*	polyglucose	reserve carbon and energy source
Polybetahydroxy butyric acid (PHB)	many bacteria e.g. *Pseudomonas*	polymerized hydroxy butyrate	reserve carbon and energy source
polyphosphate (volutin granules)	many bacteria e.g. *Corynebacterium*	linear or cyclical polymers of PO_4	reserve phosphate; possibly a reserve of high energy phosphate
sulfur globules	phototrophic purple and green sulfur bacteria and lithotrophic colorless sulfur bacteria	elemental sulfur	reserve of electrons (reducing source) in phototrophs; reserve energy source in lithotrophs
gas vesicles	aquatic bacteria especially cyanobacteria	protein hulls or shells inflated with gases	buoyancy (floatation) in the vertical water column
parasporal crystals	endospore-forming bacilli (genus *Bacillus*)	protein	toxic to certain insects
Magnetosomes	certain aquatic bacteria	magnetite (iron oxide) Fe_3O_4	orienting and migrating along geo- magnetic field lines
carboxysomes	many autotrophic bacteria	enzymes for autotrophic CO_2 fixation	site of CO_2 fixation
phycobilisomes	cyanobacteria	phycobiliproteins	light-harvesting pigments
chlorosomes	Green bacteria	lipid and protein and bacteriochlorophyll	light-harvesting pigments and antennae

4. Chromosomal DNA

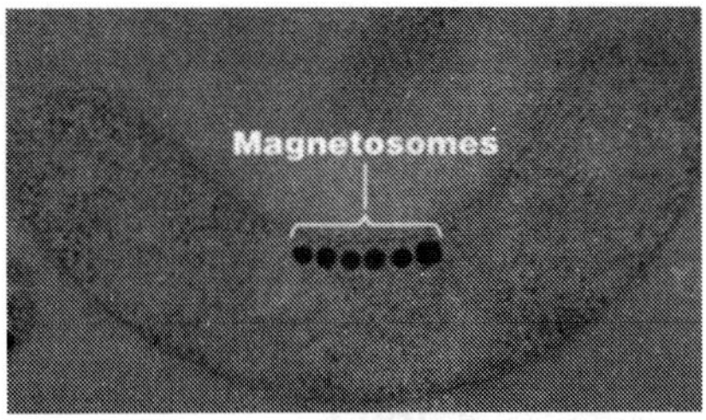

The bacteria do not contain true membrane enclosed nucleus and chromosomes. They have a long coiled double strand single circular structure often called as chromosomal DNA or nucleoid or bacterial chromosome or chromatin body.

5. Plasmids

A small circular, covalently closed, self replicating extra chromosomal DNA is also present in many bacteria. They are referred as **plasmids**. They are mostly controlling the special characters like pathogenicity, nodulation etc. *(The chromosomal DNA and plasmids are collectively called as genomic DNA).*

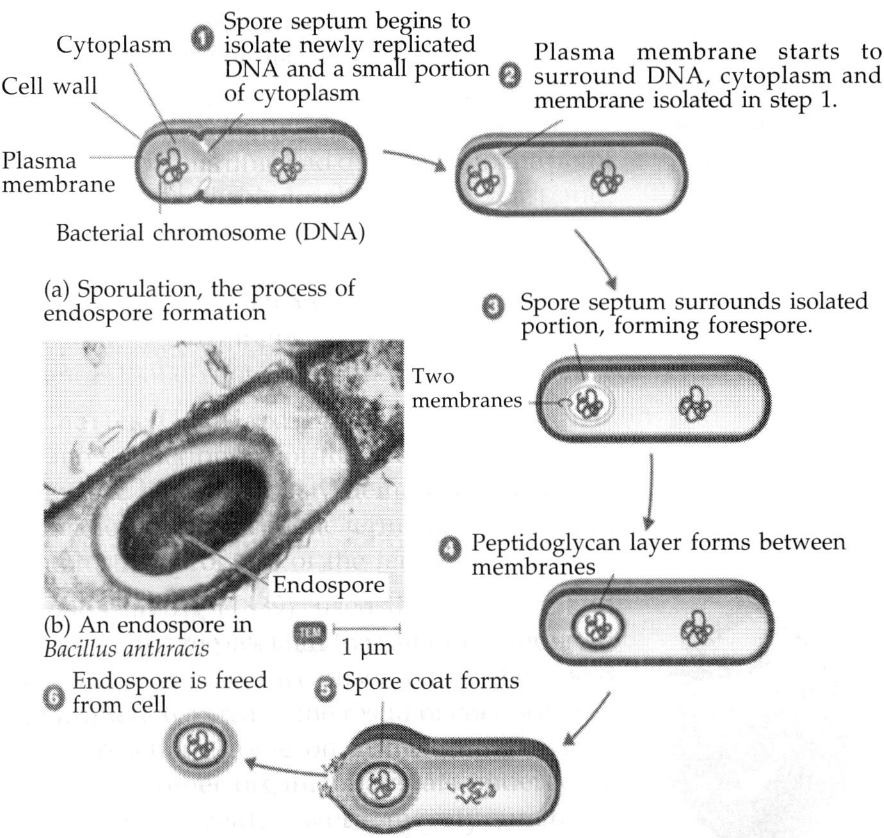

Steps involved in endospore formation

6. Endospores (Differentiation in bacteria)

Certain species of bacteria produce spores within the cells called endospores. They are unique to bacteria. *Bacillus* and *Clostridium* are common spore forming bacteria.

The endospores are extremely resistant to desiccation, staining, disinfecting chemicals, radiation and heat. The endospores are able to survive in boiling water, UV light and many harmful chemicals. They can survive even for 100 years.

One chemical substance that is characteristic of endospores and which is not available in vegetative cells is **dipicolinic acid** in the core region. They occupy 10% of total dry weight of the spores and responsible for heat resistance.

7. Cyst

Cyst is dormant, thick walled desiccation (drying) resistant form of cell. It can be differentiated from vegetative cells and can germinate under suitable condition. The cyst doesn't have temperature resistance like endospores. e.g., Cyst of *Azotobacter.*

8. Sporangiospores and Conidiospores

The filamentous bacteria, actinomycetes produce these kind of spores. The spores are produced at the tip of the hyphae and if the spores are formed in a sac like structure (Sporangium), called as sporangiospores, if not called as conidiospores. These spores do not have heat resistance but can survive long period of drying.

❑❑❑

Chapter 4

Microbial Staining

Although living microorganisms can be directly examined with the light microscope, they often must be fixed and stained to increase visibility, accentuate specific morphological features, and preserve them for future study.

Fixation

The stained cells seen in a microscope should resemble living cells as closely as possible. *Fixation* is the process by which the internal and external structures of cells and microorganisms are preserved and fixed in position. It inactivates enzymes that might disrupt cell morphology and toughens cell structures so that they do not change during staining and observation. A microorganism usually is killed and attached firmly to the microscope slide during fixation.

There are two fundamentally different types of fixation.

1. Bacteriologists heat-fix bacterial smears by gently flame heating an air-dried film of bacteria. This adequately preserves overall morphology but not structures within cells.
2. Chemical fixation must be used to protect fine cellular substructure and the morphology of larger, more delicate

microorganisms. Chemical fixatives penetrate cells and react with cellular components, usually proteins and lipids, to render them inactive, insoluble, and immobile. Common fixative mixtures contain such components as ethanol, acetic acid, mercuric chloride, formaldehyde, and glutaraldehyde.

Dyes

The many types of dyes used to stain microorganisms have two features in common. (1) They have *chromophore groups*, groups with conjugated double bonds that give the dye its color. (2) They can bind with cells by ionic, covalent, or hydrophobic bonding. For example, a positively charged dye binds to negatively charged structures on the cell. Ionizable dyes may be divided into two general classes based on the nature of their charged group.

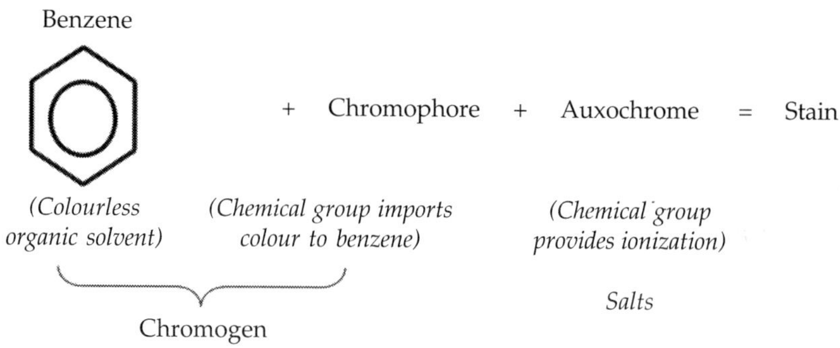

Structure of Stain

1. Basic dyes - Basic dyes are cationic. *Ex* : Methylene blue, basic fuchsin, crystal violet, safranin, malachite green. They have positively charged groups (usually some form of pentavalent nitrogen) and are generally available as chloride salts. Basic dyes bind to negatively charged molecules like nucleic acids and many proteins. Because the surfaces of bacterial cells also are negatively charged, basic dyes are most often used in bacteriology.
2. Acid dyes – Acid dyes are anionic. *Ex* : Eosin, rose bengal, and acid fuchsin. They possess negatively charged groups such as carboxyls (—COOH) and phenolic hydroxyls (—OH). Acid dyes, because of their negative charge, bind to positively charged cell structures.

The pH may alter staining effectiveness since the nature and degree of the charge on cell components change with pH. Thus anionic dyes stain best under acidic conditions when proteins and many other molecules carry a positive charge; basic dyes are most effective at higher pHs.

Although ionic interactions are probably the most common means of attachment, dyes also bind through covalent bonds or because of their solubility characteristics. For instance, DNA can be stained by the Feulgen procedure in which Schiff's reagent is covalently attached to its deoxyribose sugars after hydrochloric acid treatment. Sudan III (Sudan Black) selectively stains lipids because it is lipid soluble but will not dissolve in aqueous portions of the cell.

Types of staining

1. Simple staining

Microorganisms often can be stained very satisfactorily by simple staining, in which a single staining agent is used. Simple staining's value lies in its simplicity and ease of use. One covers the fixed smear with stain for the proper length of time, washes the excess stain off with water, and blots the slide dry. In simple staining only one stain is used. Basic dyes like crystal violet, methylene blue, and carbolfuchsin are frequently used to determine the size, shape, and arrangement of bacteria.

It is of two types:

1. *Positive staining/ Direct staining* : In which basic dyes are used. So the cells are absorbing the colour of the dye. Here cells appear coloured in bright background.
2. *Negative staining / Indirect staining* : In which acidic dyes are used. The cells will not take up the colour and instead background is stained. Here cells appear transparent against the stained or dark background.

2. Differential Staining

In differential staining more than one stain is used. The basic dyes are used. Differential staining procedures divide bacteria into separate groups based on staining properties.

Differential staining includes,

- Gram staining
- Capsule staining and
- Flagella staining

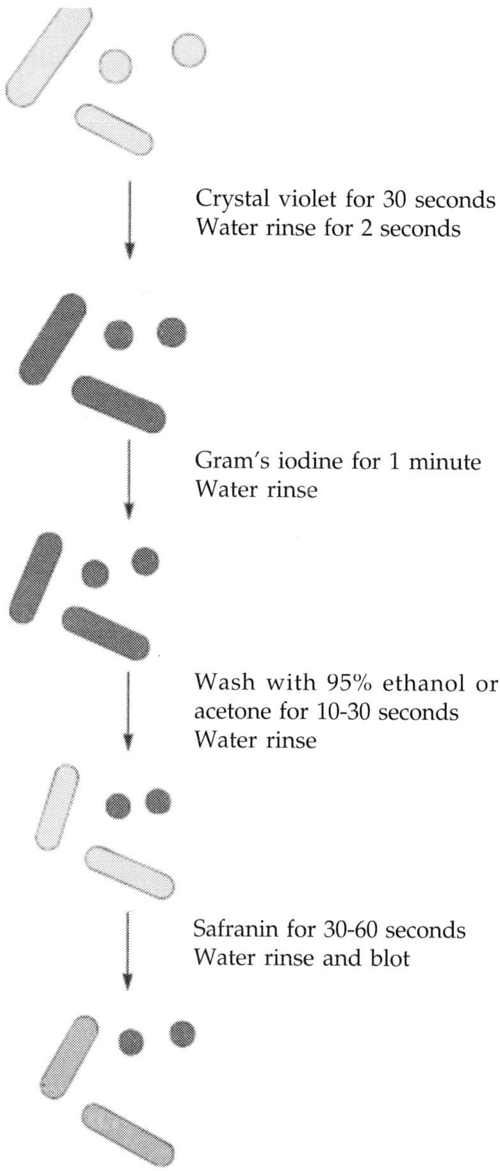

Gram reaction

The *Gram stain*, developed in 1884 by the Danish physician Christian Gram, is the most widely employed staining method in bacteriology. It is a differential staining procedure because it divides bacteria into two classes—gram negative and gram positive.

In the first step of the Gram-staining procedure, the smear is stained with the basic dye crystal violet, the primary stain. It is followed by treatment with an iodine solution functioning as a *mordant*. That is, the iodine increases the interaction between the cell and the dye so that the cell is stained more strongly. The smear is next decolorized by washing with ethanol or acetone. This step generates the differential aspect of the Gram stain; gram-positive bacteria retain the crystal violet, whereas gram-negative bacteria lose their crystal violet and become colorless. Finally, the smear is counterstained with a simple, basic dye different in color from crystal violet. Safranin, the most common counterstain, colors gram-negative bacteria pink to red and leaves gram-positive bacteria dark purple.

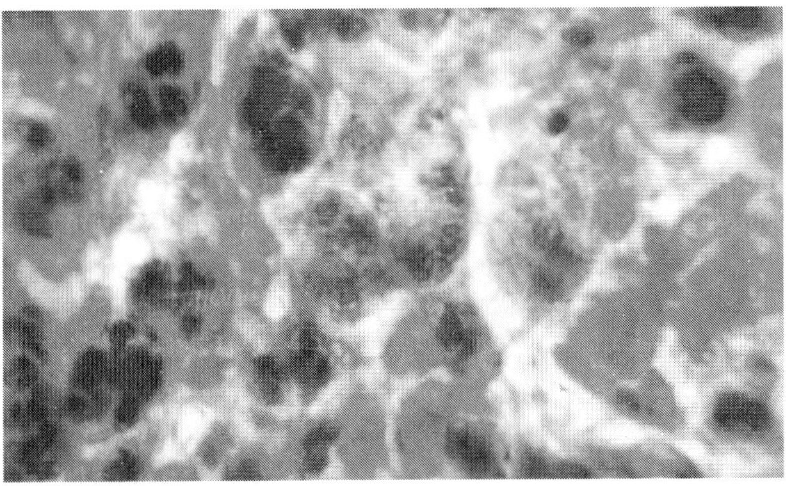

Acid-Fast Staining. *Mycobacterium leprae.* Acid-fast stain. *Note the masses of red bacteria within host cells.*

1. Acid-fast staining

It is an another important differential staining procedure. A few species, particularly those in the genus *Mycobacterium* do not bind simple stains readily and must be stained by a harsher treatment: heating with a mixture of basic fuchsin and phenol (the Ziehl-Neelsen method).

Once basic fuchsin has penetrated with the aid of heat and phenol, acid-fast cells are not easily decolorized by an acid-alcohol wash and hence remain red. This is due to the quite high lipid content of acid-fast cell walls; in particular, mycolic acid—a group of branched chain hydroxy lipids—appears responsible for acidfastness. Non-acid-fast bacteria are decolorized by acid-alcohol and thus are stained blue by methylene blue counterstain. This method is used to identify *Mycobacterium tuberculosis* and *M. leprae*, the pathogens responsible for tuberculosis and leprosy, respectively.

2. Staining Specific Structures

Many special staining procedures have been developed over the years to study specific bacterial structures with the light microscope.

a. Negative staining

One of the simplest is *negative staining*, a technique that reveals the presence of the diffuse capsules surrounding many bacteria. Bacteria are mixed with Indian ink or Nigrosin dye and spread out in a thin film on a slide. After air-drying, bacteria appear as lighter bodies in the midst of a blue-black background because ink and dye particles cannot penetrate either the bacterial cell or its capsule. The extent of the light region is determined by the size of the capsule and of the cell itself. There is little distortion of bacterial shape, and the cell can be counterstained for even greater visibility.

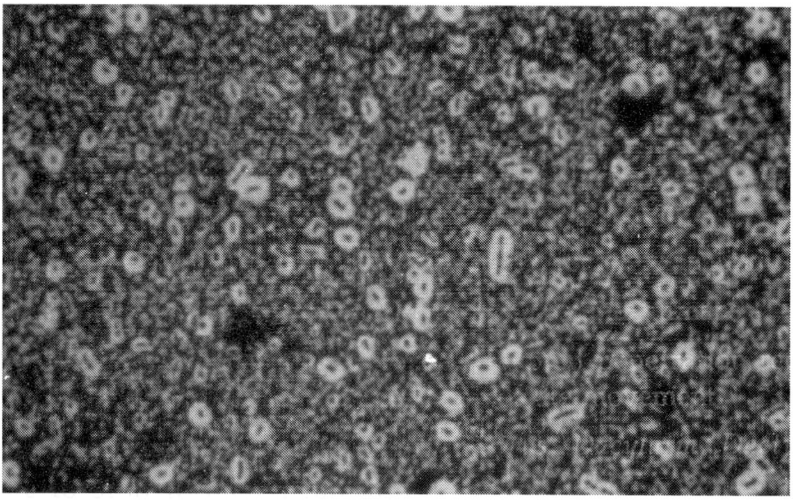

Negative Staining
Klebsiella pneumoniae negatively stained with India ink to show its capsules.

b. Endospore staining

Bacteria in the genera *Bacillus* and *Clostridium* form an exceptionally resistant structure capable of surviving for long periods in an unfavorable environment. This dormant structure is called an endospore since it develops within the cell.

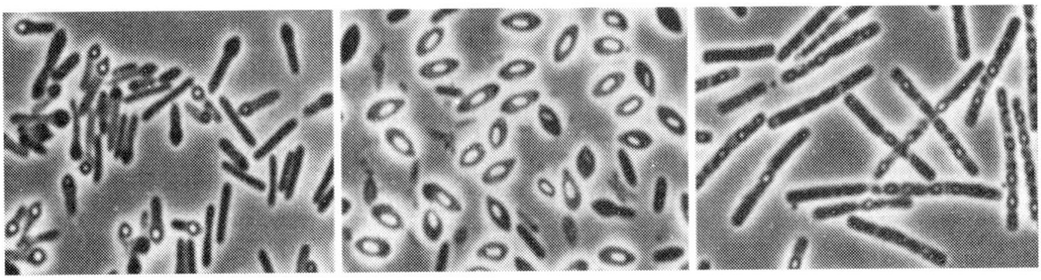

Endospore morphology and location vary with species and often are valuable in identification; endospores may be spherical to elliptical and either smaller or larger than the diameter of the parent bacterium. They can be observed with the phase-contrast microscope or negative staining. Endospores are not stained well by most dyes, but once stained, they strongly resist decolorization. This property is the basis of most *spore staining* methods.

In the Schaeffer-Fulton procedure, endospores are first stained by heating bacteria with malachite green, which is a very strong stain that can penetrate endospores. After malachite green treatment, the rest of the cell is washed free of dye with water and is counterstained with safranin. This technique yields a green endospore resting in a pink to red cell.

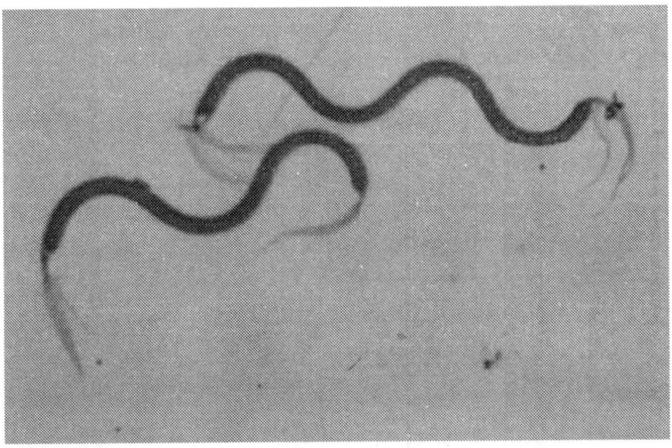

Example of Flagella Staining.
Spirillum volutans with bipolar tufts of flagella.

c. Flagella staining

Bacterial flagella are fine, threadlike organelles of locomotion that are so slender (about 10 to 30 nm in diameter) they can only be seen directly using the electron microscope. To observe them with the light microscope, the thickness of flagella is increased by coating them with mordants like tannic acid and potassium alum, and they are stained with pararosaniline (Leifson method) or basic fuchsin (Gray method). Flagella staining procedures provide taxonomically valuable information about the presence and distribution pattern of flagella.

❑❑❑

Chapter 5

Classification of Bacteria

Identification of an organism is made possible by following the classification and nomenclature guidelines and by various scientific approaches. This allows us to place an organism within its correct position in the classification scheme.

Taxonomy

Taxonomy is the science of classification, identification, and nomenclature. Taxonomy is one aspect of classification. Organisms are ordered into groups (taxa) and ranked in a hierarchy according to established procedures and guidelines. In this manner, organisms are placed into taxa of different organizational levels and the inter-relationships and boundaries between groups are established. For classification purposes, organisms are usually organized into subspecies, species, genera, families, and higher orders.

Classification

It is the orderly arrangement of bacteria into groups. There is nothing inherently scientific about classification, and different groups of scientists may classify the same organisms differently. Bacteria are classified and identified to distinguish among strains and to group them by criteria of interest to microbiologists and other scientists.

Identification

Identification is the practical use of classification criteria to distinguish certain organisms from others, to verify the authenticity or utility of a strain or a particular reaction, or to isolate and identify the organism that causes a disease.

Nomenclature

Nomenclature (naming) is the means by which the characteristics of a species are defined and communicated among microbiologists. A species name should mean the same thing to all microbiologists. Bacteria are named so that investigators can define and discuss them without the necessity of listing their characteristics. Nomenclature is another aspect of taxonomy. Names are assigned to organisms in a systematic manner.

Before scientists had a clear understanding of the nature of microbes the biological world was classified into two kingdoms: plant and animal. Bacteria were placed into plant kingdom. Clearly, this scheme was inadequate.

Haeckel suggested that a third kingdom Protista, be formed to include those unicellular microorganisms that are typically neither plants nor animals. These organisms, the protists, include bacteria, algae, fungi and protozoa. Bacteria are referred to as lower protists; the others – algae, fungi and protozoa are called higher protists.

In 1968 Whittaker proposed his famous 5 Kingdom system of living organisms. Bacteria were classified under Kingdom Prokaryotae (aka Monera). Prokaryotes were defined as "cells in which nuclear material is not surrounded by a nuclear membrane." Classification is based on the three nutritional types: photosynthesis, absorption and ingestion. The prokaryotes are included in the kingdom Monera; they lack the ingestive mode of nutrition. Unicellular eukaryotic microorganisms are included in the kingdom Protista; all three nutritional types are represented here. the mode of nutrition of microalgae is photosynthetic; the mode of nutrition of the protozoa is ingestive; and the mode of nutrition in some other protists is absorptive, with some overlap to the photosynthetic and ingestive modes. The multicellular and multinucleate eukaryotic organisms are found in the kingdom Plantae (multicellular green plants and higher algae), Animalia (multicellular animals) and Fungi (multinucleate higher fungi). Their diversified nutritional modes lead to more diversified cellular organization. Microorganisms are found in three of the five kingdoms: Monera (bacteria and cyanobacteria), Protista (microalgae and protozoa) and Fungi (yeast and mould).

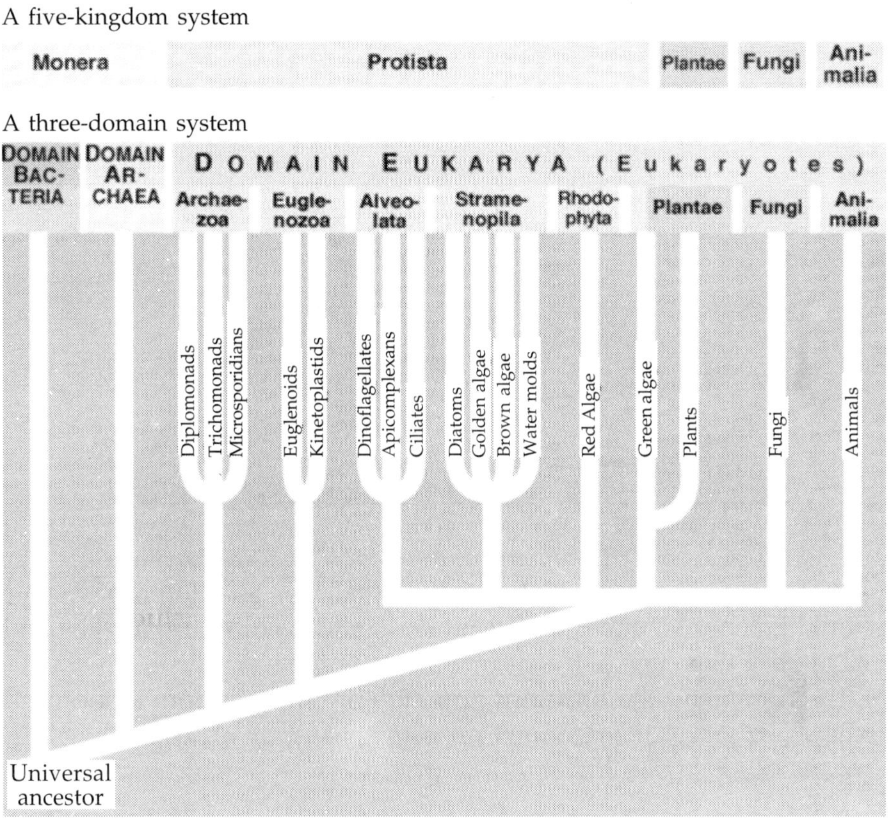

Three kingdom and five kingdom classifications

The Linnean system of Binomial Nomenclature

The Swedish naturalist, **Carolus Linneaus** developed a scientific system of naming organisms. The names used by Linneaus in the *Species Plantarum* (1753) and the *Systema Naturae* (1758) are the basis of the system for plants and animals, respectively. Linnaeus defined the biological classification system that we still use for plants and animals, and, with relatively minor modifications, for fungi and other microorganisms. It is a hierarchical system that starts with a few categories at the highest level, and further subdivides them at each lower level. The levels in the hierarchy **(taxonomical groups)** are as follows:

He assigned **two Latinized names** to each organism: A genus consists of a group of similar species. Similar genera are grouped into a family. The species name or "**specific epithet**" is unique to the new species. The genus name is indicated by a capital letter whereas the species name starts with a lower case letter. By convention both names are italicized (or underlined).

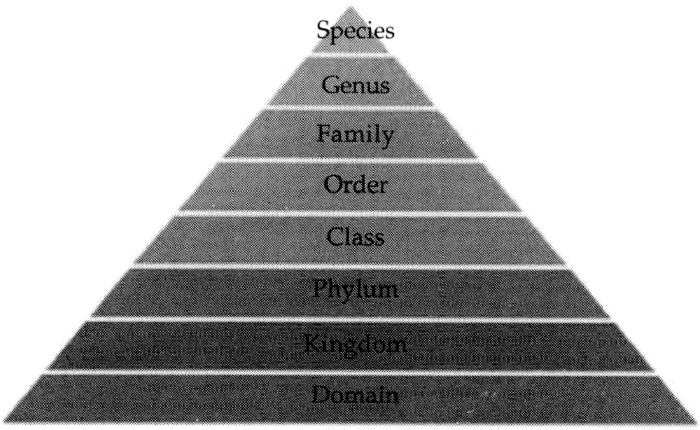

Example: ***Streptococcus pyogenes.*** This scientific name can subsequently be abbreviated as follows: ***S. pyogenes.*** Scientific names should be unique, unchanging and descriptive. For example, the name may reflect the name of the person describing the organism, the habitat of the organism, the appearance of the organism, Some names may reflect a disease or infectious process caused by an organism (e.g., 'pyogenes' describes the ability to produce pus).

A bacterial species is a distinct organism with certain characteristic features, or a group of organisms that resemble one another closely in the most important features of their organization. In the past, unfortunately, there was little agreement about these criteria or about the number of features necessary to distinguish a species. Species were often defined solely by such criteria as host range, pathogenicity, or ability to produce gas during the fermentation of a given sugar. Without a universal consensus, criteria reflected the interests of the investigators who described a particular species. For example, bacteria that caused plant diseases were often defined by the plant from which they were isolated; also, each new *Salmonella* serotype that was discovered was given species status. These practices have been replaced by generally accepted genetic criteria that can be used to define species in all groups of bacteria.

Species, groups of similar organisms within a genus, are designated by biochemical and other phenotypic criteria and by DNA relatedness, which groups strains on the basis of their overall genetic similarity.

Methods of classifying bacteria

The following methods were followed to classify bacteria:

I. Conventional method of classifying the bacteria

For identification, nomenclature and classification of any bacterium, conventionally few characters of bacteria are used. The characters are as follows:

1. *Morphological characters* : Cell shapes, size and structures; Cell arrangement; occurrence of special structures; developmental forms; staining reactions; motility and flagellar arrangement
2. *Chemical characters* : Various chemical constituents of cell like cell wall chemicals
3. *Cultural characters* : Nutritional requirement and physical conditions required for growth and the manner by which growth of bacteria occur in solid and liquid medium
4. *Metabolic characters* : The way by which the bacterium obtain their energy, carbon and electron source; regulation of these reactions
5. *Antigenic characters* : Special large chemical components (antigens) of the bacterium – distinctive for certain kind of microorganisms
6. *Genetic characters* : Characteristics of hereditary material of the cell (DNA) and occurrence and functions of other kind of DNA that may be present such as plasmid
7. *Pathogenesity* : The ability to cause the disease in various plant and animal or even other microorganisms
8. *Ecological characters* : Habitat and distribution of the organism in nature and the interaction between and among species in natural environments.

Based on these above characters, the conventional taxonomy of bacteria developed.

Subtyping

Because of differences in pathogenesity or the necessity to characterize a disease outbreak, strains of medical interest are often classified below the species level by serotyping, enzyme typing, identification of toxins or other virulence factors, or characterization of plasmids, protein patterns, or nucleic acids.

The genetic variability of microbes is further subdivided into *subspecies or types*: A *strain* is equivalent to a clone and represents a population of genetically identical organisms that have arisen from a single cell. Some strains of a bacterial species may be virulent, whereas others are not.

II. General methods of classifying bacteria

a. Phenetic system

Oganisms based on similarity of *shared phenotypic* characteristics. For example, we could place anaerobes in one group and aerobes in another. This may not always reflect the correct evolutionary groupings of the organisms. *Bergey's manual* is an example of a phenetic system. Microbes are organized into groups based on both morphological (staining reactions, cell shape and arrangement, pigment production, appearance on media) and physiological (growth requirements, biochemical tests, type of metabolism). This system can be useful for identifying an unknown organism.

b. Numerical Taxonomy

Calculates the percentage of characteristics that two organisms or groups have in common. Large range of traits (morphology, motility, biochemistry) are considered. The result of this classification is a similarity coefficient (the percentage of the total number of characters measured that are common to two organisms).

$$\text{Similarity coefficient} = \frac{\text{No. of similar characters}}{\text{No. of similar characters} + \text{No. of non similar characters}}$$

Phylogenetic system groups organisms based on their shared evolutionary heritage and descent. Unlike, a phenetic system, organisms do not have to be phenotypically similar in order to belong to the same phylogenetic group. For example, based on genetic and molecular evidence, *Pneumocystis carinii* is now considered to be more closely related to the fungi and is no longer believed to be a protozoan (although it resembles a protozoan in many respects).

III. Recent method of bacterial classification

Molecular taxonomy : Uses some key assumptions in order to establish a time-line of evolutionary relatedness. Genetic mutations are random, once a mutation occurs, all descendants of that cell will carry the mutation, organisms that differ only slightly at the genetic level have diverged more recently over the course of evolution than organisms that differ significantly.

To understand the present concept of classification, the knowledge of evolution is essential. The present bacterial classification is mainly based on the *evolutionary relationships* of the bacteria.

A brief overview on evolution

About 4.6 billion years ago, the earth was originated. Water formation occurred about 3.86 billion years ago and 3.5 billion years ago, rocks formed. Several shells of earth were formed like, *atmosphere* – $H_2$0, Nitrogen, CO_2, SO_4, CH_4; *crust* – U, K, O, Na, TH; *mantle* – Si, AL, O, Mg, Fe and *core* : Fe, Si, Ni. At that time, (3.5 billion years ago),

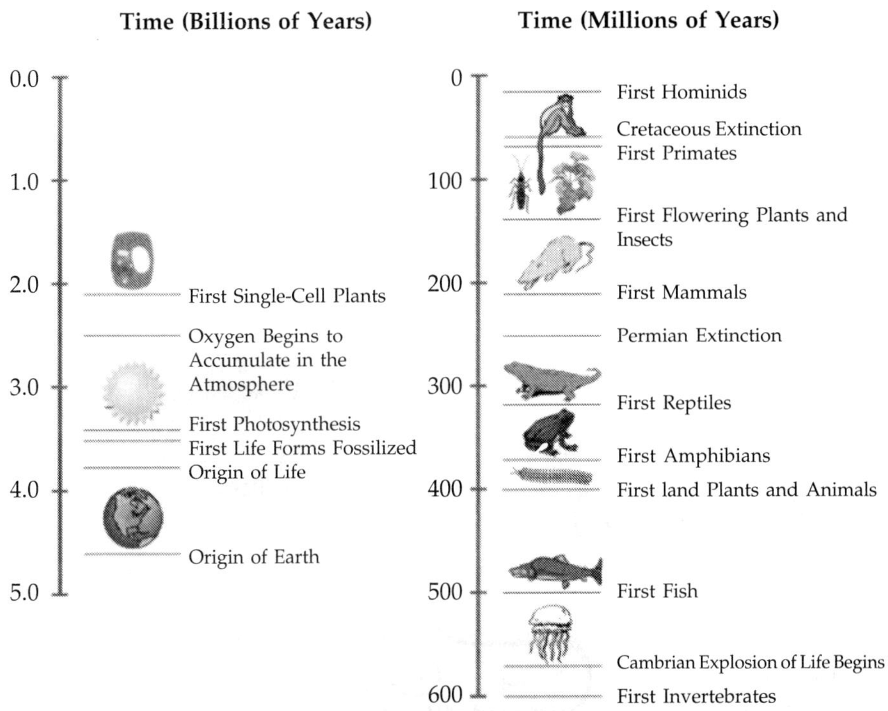

Evolution of life

in the presence of Lightening & UV lights, methane, ammonia, hydrogen, water were converted to biochemicals like amino acid, formaldehydes. Then, formed the RNA, that was able to replicate themselves. Then, RNA was surrounded by lipoprotein for the new life (RNA world). Then DNA formed and RNA acted as a coding agent. Protein played major role and thus a new life formed.

The atmosphere of that period was reduced one. The fossils are the evidence for microbial life of that period. Ancient rocks contain recognizable microfossils that appear as bacterial shapes. They are referred as *stromatolites*. These fossils were recognized as *filamentous phototrophic bacteria* (Not as present cyanobacteria – because the present one has oxygenic photophosphorylation whereas at that time the atmosphere was devoid of oxygen). The single celled phototrophs formed about 3.5 b yr ago. About 1 billion years ago, eukaryotes formed. Six hundred million years ago variety of plants and animals formed, evolved and disappeared. Mammals formed by 100 million years ago and 2-3 million years ago, MAN appeared.

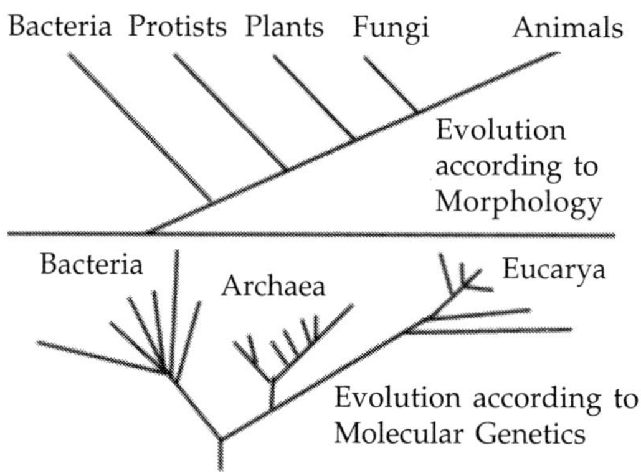

Recent method of classification of prokaryotes is mainly based on the evolutionary relationship. Molecular tools are primarily employed. Two main approaches widely used are: Comparing DNA or RNA sequences in one or more ways and comparing amino acid sequences of a protein or proteins.

There are three methods presently available for grouping the microorganisms.

- DNA : DNA hybridization
- Ribotyping (16s RNA analysis)
- FAME (Fatty acid methyl ester analysis)

Among the above three, Ribotyping is most common and frequently used tool to classify the microorganism.

a. *Hybridization of DNA probes*: The most widely used molecular method used to determine relatedness or organisms. ssDNA is separated from double stranded DNA on a filter .The DNA of one organism is radiolabelled and mixed at low concentrations with the nonradioactive denatured DNA of the other organism. The more related the organisms, the higher the degree of complementary base pairing which can be detected by a higher reading of radioactivity.

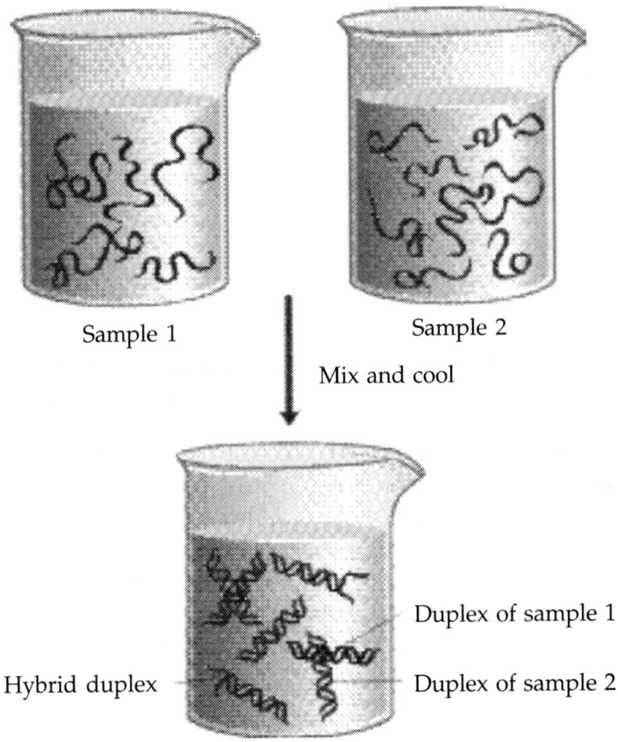

DNA : DNA hybridization

b. *16S ribosomal RNA sequences*: 16S rRNA (about 1500 nucleotides long) is found in the 30S ribosomal subunit of bacterial ribosomes. This rRNA is also found in the ribosomes of chloroplasts and mitochondria and is therefore present in animals and plants. As the ribosome plays a critical role in protein synthesis most mutations in rRNA are harmful and tend to occur very infrequently. Therefore, 16S rRNA is a very useful molecule for comparing relatedness of organisms over the course of evolution.

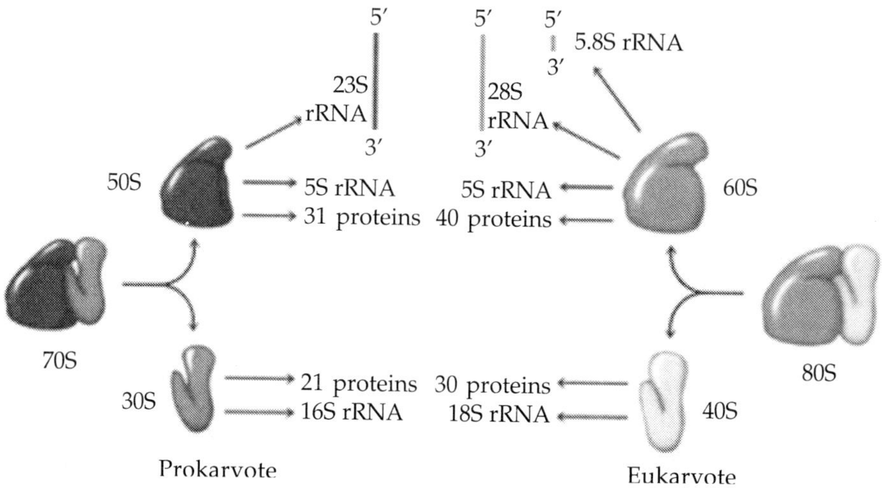

Ribosome

As the 16SrRNA is so highly conserved, organisms are classified as separate species if their sequences show less than 98% homology and are classified as different genera, if their sequences show less than 93% identity. Specific base sequences in the rRNA known as **signature sequences** were commonly found in particular groups of organisms. **Dr. Carl Woese** and colleagues at the University of Illinois examined the 16S rRNA sequences of hundreds of organisms and divided the organisms into three domains based on their signature sequences and their related properties. The domain is now the highest level of organization in the biological world.

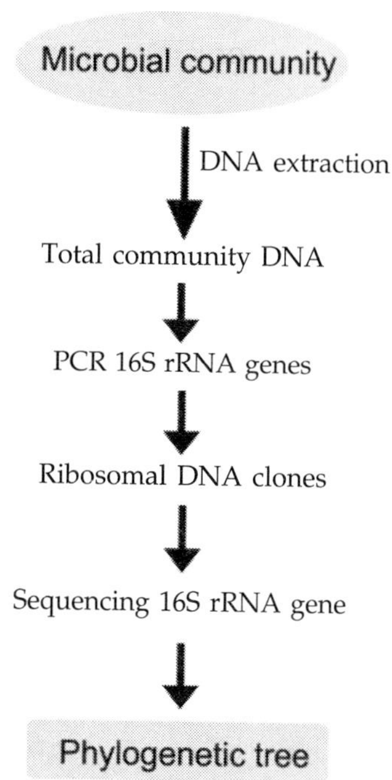

Ribosomal analysis

The Domains are termed: **Eukarya, Archaea and Eubacteria**

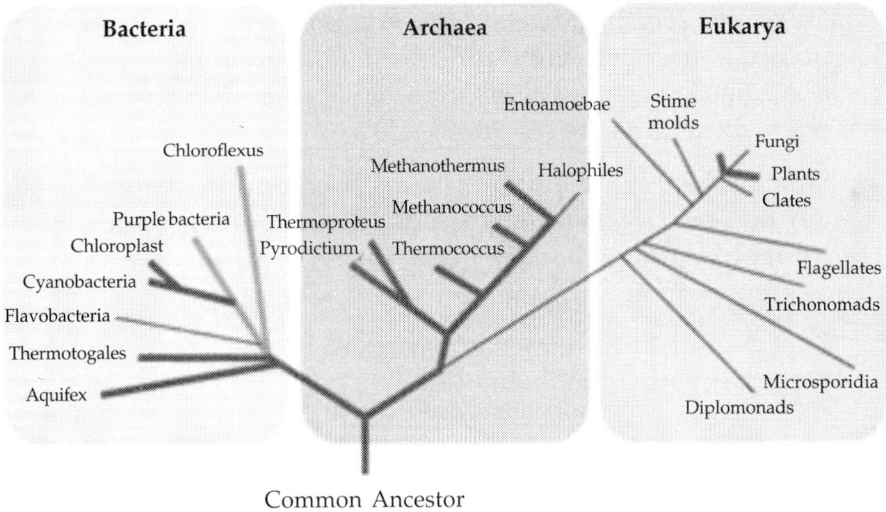

Phylogenetic tree

Among the three, two of which are prokaryotes (bacteria and archaea). This universal phylogenetic tree is sharing a common ancestor called *universal ancestor*, which is at present not known. The microorganisms fall in all the three domains.

c. *FAME* : The fatty acid composition of prokaryotes gives very high diversity. The fatty acid compositions especially the cell wall fatty acids analysis is used to identify the organisms.

d. *DNA base composition* : Indicates relatedness of organisms. Base composition is usually expressed as GC content. If the GC content differs by a small percentage the organisms are not closely related. The GC content itself does not always mean that organisms are related. For example, humans and *Bacillus* have similar GC contents but are very different organisms.

e. *DNA fingerprinting* : Comparison of the cleavage pattern (fingerprint) of the DNA from two organisms (one known, the other unknown) can determine if they are related. Each organism has a unique restriction digest profile.

f. *PCR* : Amplification and sequencing of DNA is also a powerful tool for establishing the identity of microbes and represents the diagnostic tool of the future.

IV. Other methods for identifying bacteria

a. *Direct observation*: Microscopic examination of cultures, wet mounts and staining. Since many microbes cannot be distinguished from each other on the basis of appearance alone, biochemical tests are also conducted.

b. Serological tests Identify microbes by reactivity with specific antibodies. Serotyping developed by Rebecca Lancefield. Designed A through O system to identify variants (serovars) of Streptococci.

c. Enzyme-linked immunosorbent assay (ELISA) and Immunofluorescent antibody testing (IFAT)

Universally accepted system of bacterial classification

1. Bergey's manual of determinative bacteriology

Bergey, published a first edition of Bergey's manual of determinative bacteriology from the Society of American Bacteriologists in 1923. Second edition was published in 1925, and third edition in 1930 and subsequently five editions appeared. In, 1974, the manual began a truly international cooperative effort. Authorities from all over the World were invited to prepare the descriptions of various genera and the 8th edition of the Manual released in 1974, contained contributions from 135 authors.

2. Bergey's manual of systematic bacteriology

In 1984, the scope of the Manual was greatly broadened to bring information dealing with ecology, environment, isolation, perversion & characterization of bacteria all of which concerned with bacterial classification and identification. The new breath of coverage was reflected by the new name 'Bergey's Manual of Systematic Bacteriology'. As the knowledge about bacteria continues to increase, so will Bergey's manual continue to change with it and to act as a "Mirror" for the World of bacteria. Bergey's manual is the international standard for bacterial taxonomy and nomenclature.

Bergey's manual of systematic bacteriology places all bacteria in the kingdom "Prokaryote" which in turn is divided into 4 divisions.

Division: 1 Gracilicutes: Bacteria with a thin cell wall. Gram –ve bacteria are included in this division

Division: 2 Firmicutes: Bacteria with a thick cell wall structure. (characteristic ofGram +ve bacteria)

Division: 3 Tenericutes: Bacteria with tender and soft cell wall. Mycoplasmas are included in this division.

Division: 4 Mendosicutes: Procaryotes that show evidence of an earlier phylogenetic origin than those bacteria included in Div.1 and Div.2. They have faulty (mendose=faulty) cell walls. They do not have typical peptidoglycon, instead they have proteins. Also they live in unusual habitats like hot springs.

Bergey's manual of systematic bacteriology is a compendium of standard and molecular informations about the available prokaryotes.

First edition consists of 4 volumes

Vol 1. Gram positive bacteria

Vol 2. Gram negative bacteria

Vol 3. Bacteria with unusual properties including archaea

Vol 4. Filamentous bacteria

This system does not have much phylogenic informations. The second edition of the Bergey's manual provides much clear informations about the genetic relationship among the organisms.

The second edition was divided into 5 volumes

Vol 1. Archaea, cyanobacteria, phototrophs and deeply branched genera

Vol 2. Proteobacteria

Vol 3. Low G+C gram positives

Vol 4. High G+C gram positives

Vol 5. Planctomycetes, Spirochetes, Bacteroides, Fusobacteria

The examples of each group of bacteria are as follows:

Vol 1. Archaea, cyanobacteria, phototrophs and deeply branched genera

This volume has 3 important groups out of which one is in different domain (Domain–archaea)

a. Archaea

Four important sections are present in the archaea.

1. Hyperthermophiles Ex. *Thermococcus, Sulfolobus, Thermosphaera*
2. Methanogens Ex. *Methanobacterium, Methanococcus, Methanosarcina*
3. Halobacteria Ex. *Halobacterium, Halococcus, Natronomonas*
4. Thermoplasma Ex. *Thermoplasma*

b. Cyanobacteria

They are the filamentous, oxygenic photosynthetic bacteria. They have special cells called heterocyst in which nitrogenase enzyme is present. The nitrogenase enzyme is responsible for fixing atmospheric N_2 into ammonia. Some cyanobacteria are single celled and some are non-heterocystous. Important groups are as follows:

1. Single celled cyanobacteria Ex. *Chrococcus, Gleotheca, Gleocapsa*
2. Filamentous non-heterocystous cyanobacteria Ex. *Oscillatoria, Lyngbya*
3. Filamentous heterocystous cyanobacteria Ex. *Anabaena, Nostoc, Tolypothrix*

c. Anoxygenic phototrophs

Single celled, sulphur requiring bacteria. They use H_2S as electron donor. Ex. Green sulphur bacterium *Chlorobium*.

Bergey's Manual - Volume I

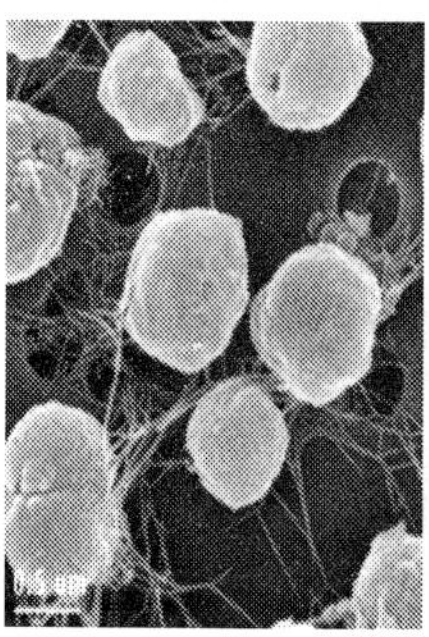

Methanococcus

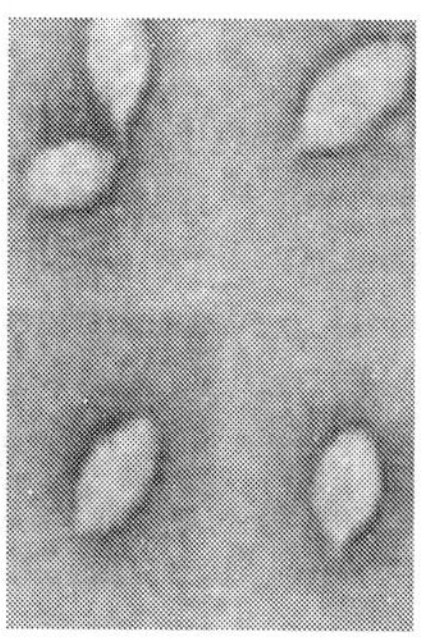

Sulfolobus

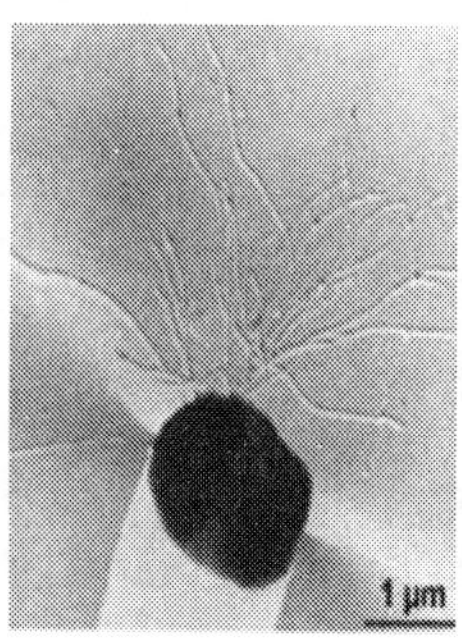

Pyrococcus

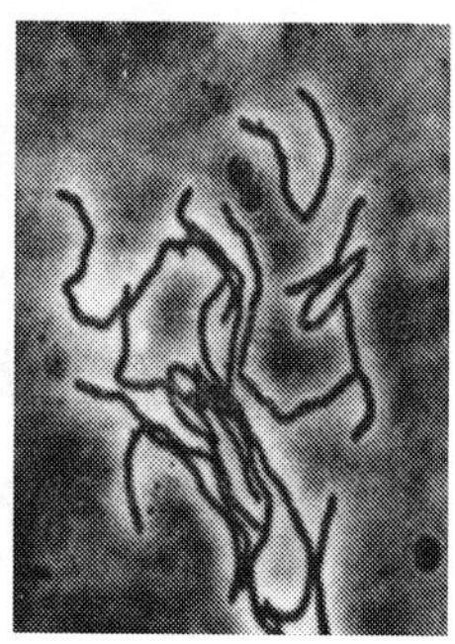

Methanobrevibacter

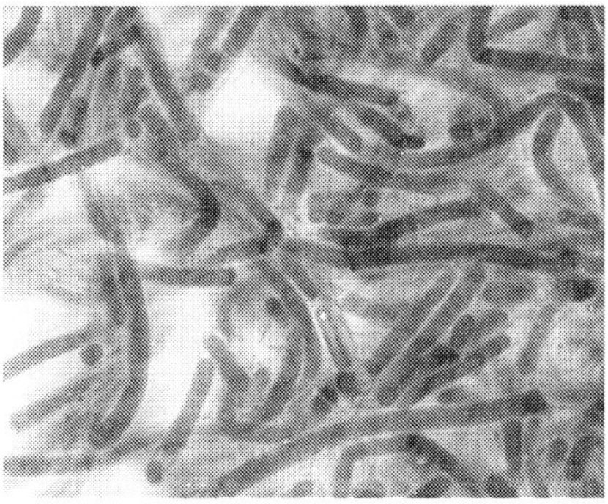

Cyanobacteria

Vol 2. Proteobacteria

This volume has gram negative bacteria. They are further divided into 5 subgroups as a, b, g, d and e. They contain medically, industrially and agriculturally important bacteria.

S. No.	Important bacteria	Characters	Example
α ***Proteobacteria***			
1.	Purple bacteria	Anoxygenic Photosynthetic – sulphur bacteria	*Rhodospirillum, Rhodobacter, Chromatium*
2.	Associative Nitrogen fixing bacteria	These bacteria present in the rhizosphere of graminaceous plants and symbiotically fix atmospheric nitrogen.	*Azospirillum*
3.	Symbiotic Nitrogen fixing bacteria	Form nodules in legume roots and fix atmospheric nitrogen.	*Rhizobium, Bradyrhizobium,*
		Some form galls in the roots	*Agrobacterium*
4.	Free living Nitrogen fixing bacteria	Present in the soil as heterotrophs – use verity of carbon sources in soil and fix atmospheric nitrogen	*Azotobacter, Beijerinkia*
5.	Pseudomonas group	Some are Plant Growth Promoting Rhizobacteria	*Pseudomonas*
		Some are pathogens	*Xanthomonas*
		Some produce alcohol	*Zymomonas*
6.	Rickettsia	Endoparasites	*Rickettsia*

7.	Sulphur oxidizing bacteria	Uses S as electron donor – Chemolithotrophs – strict aerobes	*Thiobacillus*
8.	Acetic acid producing bacteria	Fermentative bacteria	*Acetobacter, Gluconobacter*
9.	Budding bacteria	Reproduction by budding like yeast	*Caulobacter*
10.	Hydrogen bacteria	Hydrogen producing bacteria	*Alkaligenes*
β *Proteobacteria*			
1.	Nitrifying bacteria	Chemolithotroph – strict aerobe – soil bacteria – important form N cycle	*Ammonia to nitrite – Nitrosomonas Nitrite to nitrate – Nitrobacter*
2.	Neisseria & relatives		*Neisseria*
3.	Spirillum	Aerobes & facultative aerobes	*Spirillum sp.*
4.	Sheathed bacteria		*Sphaerotilus*
γ *Proteobacteria*			
1.	Purple sulphur bacteria	Anoxygenic photosynthetic – sulphur bacteria	*Thiobacillus, Thiospirillum*
2.	Methylotrophs	Uses methane and methanol as carbon source	*Methylomonas, Methylobacter, methylococcus*
3.	Coliforms	Present in the intestinal track of mammals	*Escherichia, Salmonella*
δ *Proteobacteria*			
1.	Sulphur reducing bacteria	Anaerobes – use S as terminal electron acceptor	*Desulfovibrio, Desulfomonas*
2.	Gliding bacteria	Gliding movement	*Myxobacteria*
3.	Vibrio group	Most are pathogenic	*Vibrio, Erwinia*
ε *Proteobacteria*			
1.	*Camphylobacter, Thiovulum, Thiomicrospora, Heliobacter* are common examples		

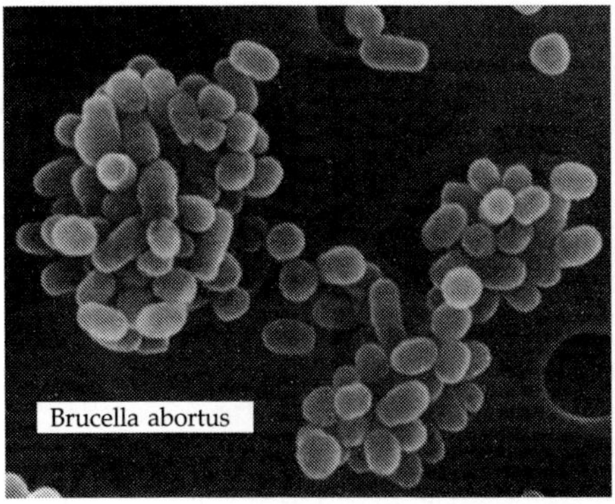

Brucella abortus

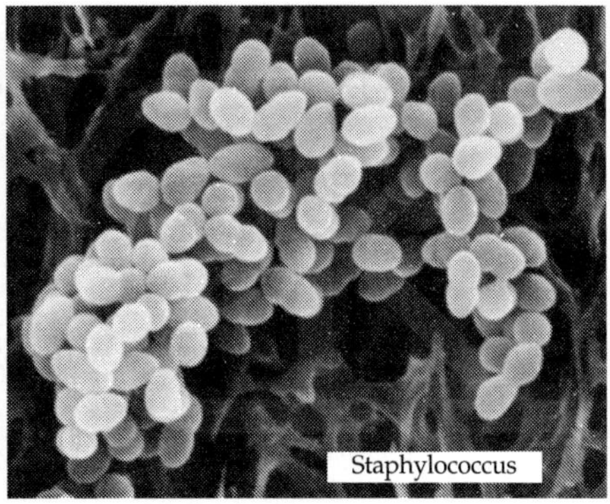

Staphylococcus

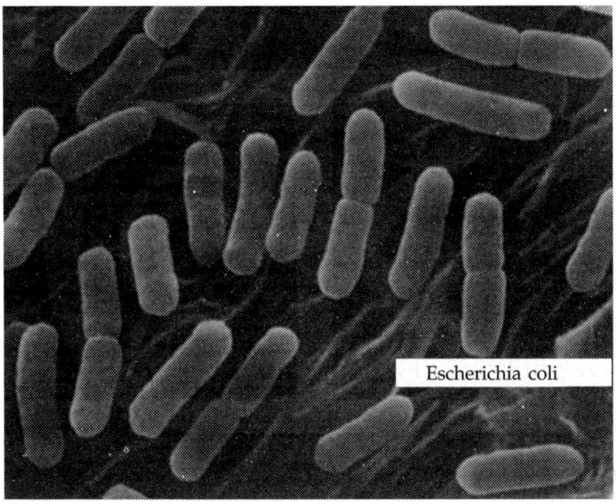

Escherichia coli

Bergey's Manual - Volume II

Vol 3. Low G+C gram positives

S. No.	Group	Characters	Example
1.	Clostridia group	Strict anaerobes – mostly fermentative nutrition – few thermotolerant – endospore producers	*Clostridium, Thermoanaerobacterium, Thermoanaerobium*
2.	Mycoplasma group	Absence of cell wall	*Mycoplasma, Mesoplasma, Spiroplasma*
3.	Bacilli and Lactobacilli group	Lactic acid producing bacteria – endospore producers – aerobes – aerotolerant – fermentative nutrition	*Leuconostoc, Lactococcus, Streptococcus*

Vol 4. High G+C gram positives

S. No.	Group	Characters	Example
1.	Actinomycetes	Filamentous – sporangiospores – conidiospores – soil habitat – antibiotics producers	*Actinomyces, Nocardia, Sreptomyces*
		Symbiotic with *Casuarina* – form root nodules – N_2 fixation	*Frankia*
2.	Mycobacterium group	Presence of mycolic acid in the cell wall – acid fast staining – human pathogens	*Mycobacterium lepri*
3.	Corynebacterium group	Human pathogens	*Corynebacterium diptheriaea*

Bergey's Manual - Volume III and IV

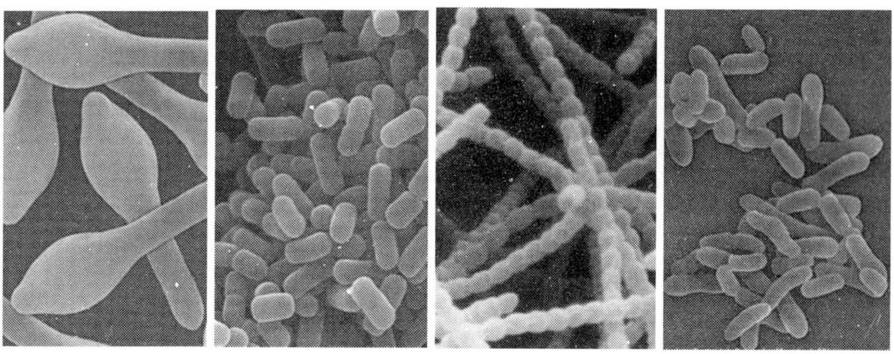

Clostridium *Lactobacillus* *Streptomyces* *Mycobacterium*

Vol 5. Plancomycetes, Spirocheter, Bacteroides and Fusobacteria

S. No.	Group	Characters	Example
1.	Chlamydia group	Obligate parasites to man, animal and birds	*Chlamydia*
2.	Bacteroides	Obligate anaerobes	*Bacteroides*
3.	Spirochete	Gram negative – flexile – endoflagella presence	*Spirocheta, Leptospira*

❑❑❑

Chapter 6

Nutritional Types of Bacteria

All forms of life, from microorganisms to human beings, share certain nutritional requirements for growth and normal functioning. This chapter substantiates this statement and also illustrates the great diversity of nutritional type found among bacteria.

Requirements for Carbon, Hydrogen and Oxygen

The requirements for carbon, hydrogen, and oxygen often are satisfied together. Carbon is needed for the skeleton or backbone of all organic molecules, and molecules serving as carbon sources normally also contribute both oxygen and hydrogen atoms. They are the source of all three elements. Because these organic nutrients are almost always reduced and have electrons that they can donate to other molecules, they also can serve as energy sources.

Indeed, the more reduced organic molecules are, the higher in their energy content (e.g., lipids have higher energy content than carbohydrates). This is because, as we shall see later, electron transfers release energy when the electrons move from reduced donors with more negative reduction potentials to oxidized electron acceptors with more positive potentials. Thus carbon sources frequently also serve as energy sources.

One important carbon source that does not supply hydrogen or energy is carbon dioxide (CO_2). This is because CO_2 is oxidized and lacks hydrogen. Probably all microorganisms can fix CO_2 - that is, reduce it and incorporate it into organic molecules. However, by definition, only *autotrophs* can use CO_2 as their sole or principal source of carbon. Many microorganisms are autotrophic, and most of these carry out photosynthesis and use light as their energy source. Some autotrophs oxidize inorganic molecules and derive energy from electron transfers.

The reduction of CO_2 is a very energy-expensive process. Thus many microorganisms cannot use CO_2 as their sole carbon source but must rely on the presence of more reduced, complex molecules such as glucose for a supply of carbon. Organisms that use reduced, preformed organic molecules as carbon sources are *heterotrophs* (these preformed molecules normally come from other organisms). As mentioned previously, most heterotrophs use reduced organic compounds as sources of both carbon and energy. For example, the glycolytic pathway produces carbon skeletons for use in biosynthesis and also releases energy as ATP and NADH.

A most remarkable nutritional characteristic of microorganisms is their extraordinary flexibility with respect to carbon sources. Laboratory experiments indicate that there is no naturally occurring organic molecule that cannot be used by some microorganism. Actinomycetes will degrade amyl alcohol, paraffin, and even rubber. Some bacteria seem able to employ almost anything as a carbon source; for example, *Burkholderia cepacia* can use over 100 different carbon compounds. In contrast to these bacterial omnivores, some bacteria are exceedingly fastidious and catabolize only a few carbon compounds. Cultures of methylotrophic bacteria metabolize methane, methanol, carbon monoxide, formic acid, and related one-carbon molecules. Parasitic members of the genus *Leptospira* use only long-chain fatty acids as their major source of carbon and energy.

It appears that in natural environments complex populations of microorganisms often will metabolize even relatively indigestible human-made substances such as pesticides. Indigestible molecules sometimes are oxidized and degraded in the presence of a growthpromoting nutrient that is metabolized at the same time, a process called *cometabolism*. The products of this breakdown process can then be used as nutrients by other microorganisms.

Nutritional Types of Microorganisms

In addition to the need for carbon, hydrogen, and oxygen, all organisms require sources of energy and electrons for growth to take place. Microorganisms can be grouped into nutritional classes based on how they satisfy all these requirements.

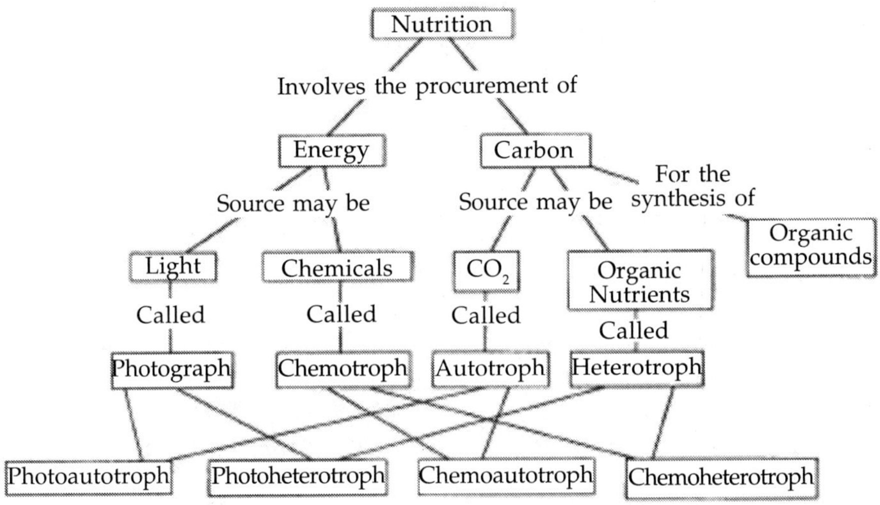

We have already seen that microorganisms can be classified as either heterotrophs or autotrophs with respect to their preferred source of carbon.

There are only two sources of energy available to organisms:

1. light energy, and
2. the energy derived from oxidizing organic or inorganic molecules.

Phototrophs use light as their energy source; chemotrophs obtain energy from the oxidation of chemical compounds (either organic or inorganic). Microorganisms also have only two sources for electrons. Lithotrophs (i.e., "rock-eaters") use reduced inorganic substances as their electron source, whereas organotrophs extract electrons from organic compounds.

Despite the great metabolic diversity seen in microorganisms, most may be placed in one of four nutritional classes based on their primary sources of carbon, energy, and electrons. The large majority of microorganisms thus far studied are either photolithotrophic autotrophs or chemoorganotrophic heterotrophs.

Photolithotrophic autotrophs (often called *photoautotrophs* or photolithoautotrophs) use light energy and have CO_2 as their carbon source. Eucaryotic algae and cyanobacteria employ water as the electron donor and release oxygen. Purple and green sulfur bacteria cannot oxidize water but extract electrons from inorganic donors like hydrogen, hydrogen sulfide, and elemental sulfur.

Chemoorganotrophic heterotrophs (often called *chemoheterotrophs*, chemoorganoheterotrophs, or even heterotrophs) use organic compounds as sources of energy, hydrogen, electrons, and carbon. Frequently the same organic nutrient will satisfy all these requirements. It should be noted that essentially all pathogenic microorganisms are chemoheterotrophs.

The other two nutritional classes have fewer microorganisms but often are very important ecologically. Some purple and green bacteria are photosynthetic and use organic matter as their electron donor and carbon source. These *photoorganotrophic heterotrophs* (photoorganoheterotrophs) are common inhabitants of polluted lakes and streams. Some of these bacteria also can grow as photoautotrophs with molecular hydrogen as an electron donor.

The fourth group, the *chemolithotrophic autotrophs* (chemolithoautotrophs), oxidizes reduced inorganic compounds such as iron, nitrogen, or sulfur molecules to derive both energy and electrons for biosynthesis. Carbon dioxide is the carbon source. A few chemolithotrophs can derive their carbon from organic sources and thus are heterotrophic. Chemolithotrophs contribute greatly to the chemical transformations of elements (e.g., the conversion of ammonia to nitrate or sulfur to sulfate) that continually occur in the ecosystem.

Although a particular species usually belongs in only one of the four nutritional classes, some show great metabolic flexibility and alter their metabolic patterns in response to environmental changes. For example, many purple nonsulfur bacteria act as photoorganotrophic heterotrophs in the absence of oxygen but oxidize organic molecules and function chemotrophically at normal oxygen levels. When oxygen is low, photosynthesis and oxidative metabolism may function simultaneously. Another example is provided by bacteria such as *Beggiatoa* that rely on inorganic energy sources and organic (or sometimes CO_2) carbon sources. These microbes are sometimes called *mixotrophic* because they combine chemolithoautotrophic and heterotrophic metabolic processes. This sort of flexibility seems complex and confusing, yet it gives its possessor a definite advantage if environmental conditions frequently change.

Overview of different nutritional types of bacteria as combinations

Groups	Energy	Electron	Carbon source	Example
Photolithotroph	Sunlight	Inorganic compounds	Inorganic compounds (CO_2)	*Chromatium* (H_2S, CO_2)
Chemolithotroph	Chemicals (organic and inorganic)	Inorganic compounds	Inorganic compounds (CO_2)	*Nitrobacter, Nitrosomonas* (NH_3,NO_2, Glucose,CO_2)
Photoorganotroph	Sunlight	organic compounds	organic compounds	*Rhodospirillum* (Purple S) (Light, fatty acids)
Chemoorganotroph	Chemicals (organic and inorganic)	organic compounds	organic compounds	*E.coli* (Glucose)
Photoautotroph	Sunlight	Inorganic compounds	Inorganic compounds (CO_2)	Cyanobacteria, *Chromatium* (light,CO_2)
Photoheterotroph	Sunlight	organic compounds	organic compounds	*Rhodospirillum* (Light, Succinate)
Chemoautotroph	Chemicals (organic and inorganic)	Inorganic compounds	Inorganic compounds (CO_2)	*Thiobacillus* (sulphide, CO_2)
Chemoheterotrophs	Chemicals (organic and inorganic)	organic compounds	organic compounds	*E.coli*

❑❑❑

Chapter 7

Bacterial Growth

Previous chapter emphasized more that, microorganisms are of different nutritional types and are highly diverse. This chapter considerably explores the growth aspects of bacteria.

A. Terminologies in bacterial growth

Growth

Growth may be defined as an increase in cellular constituents. It leads to a rise in cell number when microorganisms reproduce by processes like budding or binary fission. In the latter, individual cells enlarge and divide to yield two progeny of approximately equal size. Growth also results when cells simply become longer or larger. If the microorganism is coenocytic - that is, a multinucleate organism in which nuclear divisions are not accompanied by cell divisions, growth results in an increase in cell size but not cell number. It is usually not convenient to investigate the growth and reproduction of individual microorganisms because of their small size. Therefore, when studying growth, microbiologists normally follow changes in the total population number.

Step wise reactions occur during the growth of bacteria are as follows:

- The macromolecules such as DNA, RNA, protein, and polysaccharides are synthesized
- Assembly of the synthesized macromolecules
- Cell organelles like cell wall, membrane, ribosome, etc. formation occur

Binary fission

Most bacteria divide by a process known as *binary fission*. Prokaryotes lack a mitotic apparatus. Cell elongation followed by DNA replication occurs. DNA is replicated bidirectionally and then is allocated into two new, genetically daughter cells. Cells may remain attached forming characteristic packets, chains or filaments. During the cell division mother cell divides into 2 daughter cells.

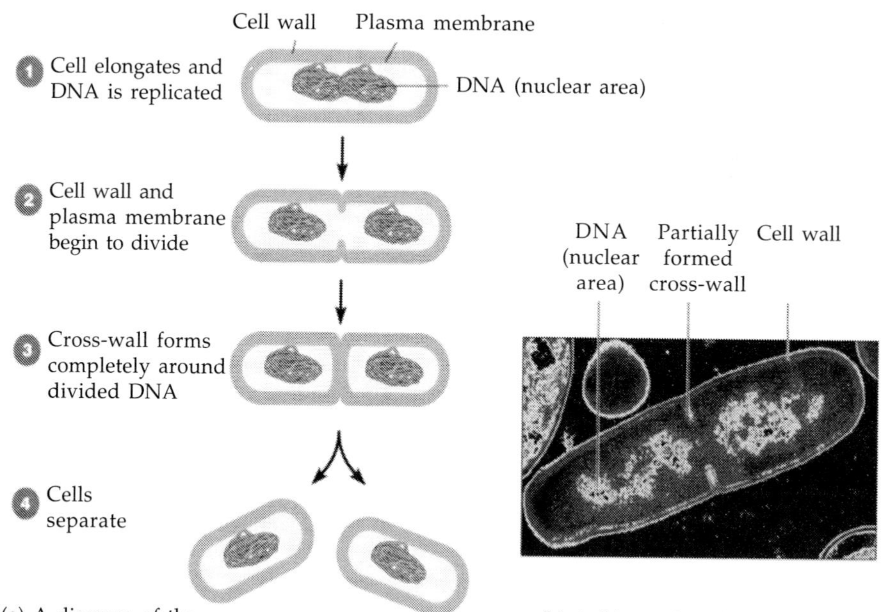

(a) A diagram of the sequence of cell division.

(b) A thin section of a cell of *Bacillus licheniformis* starting to divide.

Growth rate

Increase in bacterial cell number per unit time is referred as growth rate

Generation

The interval of formation of daughter cells from parental cell is called as generation

Generation time

Time required to form two daughter cells from a single cell is called as generation time. This is also called as doubling time. *E. coli* has the doubling time of 20 min and *Rhizobium* has 2 h.

B. Growth curve of bacteria

Assuming a single bacterium has been inoculated to a liquid medium and the bacterium starts multiplying by binary fission gives increase of population. If we plot a graph of $\log_{10}$ value of population against time, the graph will be as follows:

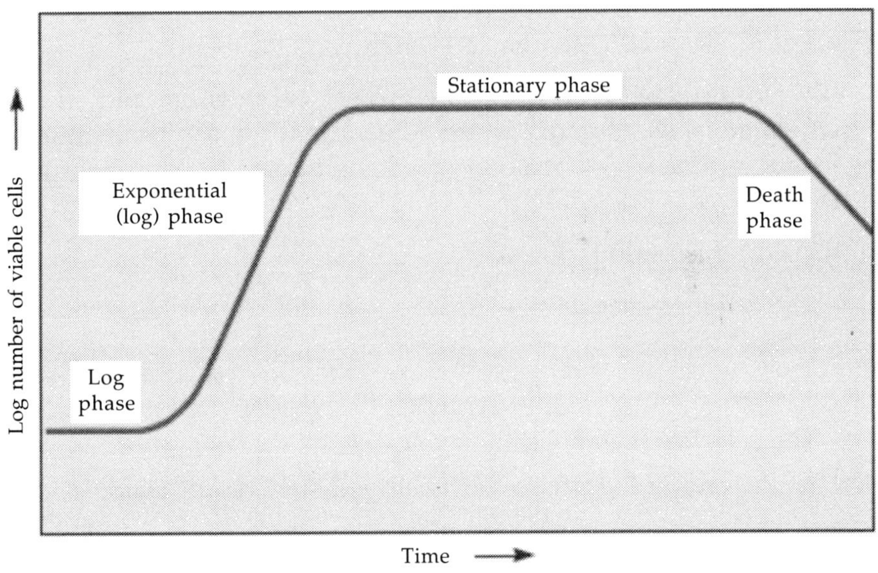

This curve is referred as growth curve of bacteria. The curve can be divided in to 4 phases. In the beginning, the growth (increasing the population) will not occur and it leads to straight line (called *lag phase*), then the cells start multiplying and form linear line (*log or exponential phase*) followed by straight line due to balance in growth & death (called *stationary phase*) and decline linear line due to continuous death (*death phase or decline phase*).

Phases of bacterial growth

1. Lag phase

When a bacterium is introduced into a medium, it will not multiply immediately. After a period of time only the bacterium will start to multiply. This period is called as lag phase.

Lag is lack in cell division but not lack in synthesis of new protoplasm. Period where cells don't increase in number but are metabolically active and increasing in size- synthesizing structural components DNA, ribosome and enzymes to breakdown nutrients, replicating the chromosome, taking in and storing nutrients to be used for growth, expelling wastes, etc. Cytochemistry reveals an increase in protein content per cell but not in proportion to nucleic acids. RNA synthesis is accelerated and varies with the length of generation time. DNA synthesis is not affected and DNA per cell is constant. Cell size may increase to 2 to 3 times and even more. Although part of the increase is due to higher water content. The increase in size is more apparent in all rod shaped bacteria than cocci.

Bacterium may be busy but the growth curve is flat and all this activity is invisible. Organisms are adjusting to the environment and the length of time varies by species and conditions- from an hour or 2 to several days.

During this period,

- The cell increases its size
- It will synthesis new protoplasm, enzymes, co enzymes etc
- The bacterium will adjust to the physiological environment
- The cells are metabolically active
- At the end of lag phase, each cell will start reproduction

2. Log / exponential phase

After lag phase, the cells start to multiply steadily by binary fission at constant rate.

Log Phase is also called as *period of exponential growth.* During this phase one cell divides into 2, those 2 into 4, and these 4 into 16 etc. Log phase requires optimal conditions meaning whatever the bacteria need for nutrients must be available and the temperature must be right. Division is at a constant rate (generation time) but varies with species, temperature and media. Cells are most susceptible to inhibitors.

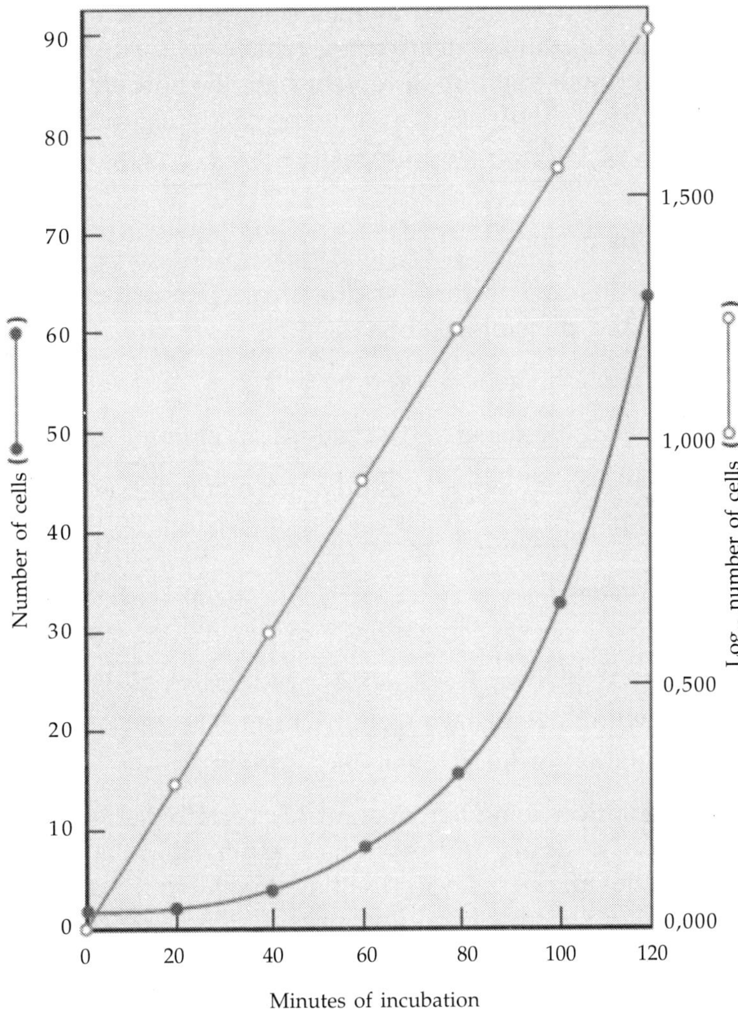

Exponential Microbial Growth

The number of bacteria in a population in log phase can be calculated, if the starting number and the generation time are known. The formula is:

Nf = Ni x 2^n

Where,
n = the number of generations
Nf = the final number in a population and
Ni = the initial number in a population

Nutrients get used up and as they diminish, growth slows. Also this heavy metabolism creates wastes, which are toxic or poisonous, and these toxins will build up. If replenishing the nutrients and getting rid of the wastes is there, then log could be continued indefinitely. The portion of curve is linear and have constant generation or doubling time.

During this period,

- All the cells are uniform in chemical composition, metabolic activity and physiological characters
- At this time, the generation time will be of constant
- The cells will be sensitive to chemical inhibitors
- This is the ideal stage to study the metabolic activity of the bacterium

3. Stationary phase

Following log phase, generations and growth rate starts to decline as a consequence of

a) Approaching exhaustion of nutrients
b) Accumulation of toxic metabolic products

The population remains constant for certain period of time as during this phase dying and dividing organisms are at equilibrium meaning that the number of cells dividing is equal to the number dying and it looks like no growth on the curve. Cells are smaller and have fewer ribosomes. In some cases cells do not die but they are not multiplying. The nutrients continue to diminish, the wastes to build, in addition to pH changes, toxic waste and reduced oxygen.

During this stage,

- Cell division rate and cell death rate are uniform
- Death due to low nutrients content, pH change, toxic wastes, reduced oxygen occur
- Cells are very small and having low number of ribosomes
- Some cases, cell do not die and at the same time do not multiply
- Some bacteria produce antibiotics at this stage as secondary metabolites for competitiveness

4. Death phase

In this phase, few cells are alive and more cells are dead. The death rate is higher than the growth rate which leads to decline in the population. This phase is referred as death phase or *decline phase.*

During this stage,

- The cell lysis occurs
- Some species can die very rapidly and some can persist for months

Note : These growth curve and the terms lag, log, stationary and decline phases are used only for the population or group of bacterial cells not for individual cells.

C. Synchronous growth of bacteria

All the cells of a bacterial culture in same stage of their growth cycle is referred as synchronous growth.

When bacterial growth which we discussed earlier as growth curve, if observed under microscope, will have different stages of its reproduction ie, cell elongation, septum formation, divided cells etc. So, the cells and cell divisions will not be uniform. But, in synchronous growth culture, the bacterial cells will be of uniform stage.

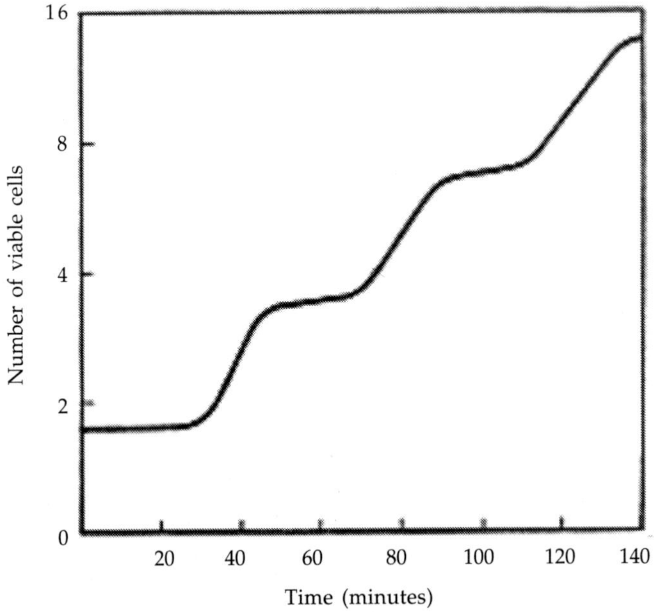

The synchronous growth of bacteria can be achieved by growing the cells at sub optimal temperature (below the optimum temperature). E. coli needs 37°C for its normal growth and if grown at 28°C, the synchronous growth will occur. This type of culture will be useful in molecular biology especially the gene cloning and transformation studies.

D. Steady state growth of bacteria

Maintaining the bacterial population in a particular stage / phase of the cycle is referred as study state of growth. The growth curve which discussed earlier is meant for the bacterial culture in a constant media referred as *batch culture*. In this type of culture, the maintenance of bacteria at particular stage is not possible. For achieving the study state growth, the constant population should be maintained in the medium. This can be achieved by adding new medium and removing few cells at frequent time interval. This type of culture is referred as *continuous culture*. The bacterium at late log phase is much interested for many industrial processes. So, maintaining the bacteria at log phase can be achieved by two systems called *chemostat* and *turbidostat*.

1. *Chemostat*: Maintaining the population of bacteria at particular stage using an important nutrient source as growth control is referred as chemostat. Ex. In a medium, if the glucose concentration is maintained by subsequent addition at particular dilution rate, so that it will be taken up by the organism immediately and at the same time a portion of the culture will be removed subsequently. This condition will maintain the bacterial growth at particular stage especially in log phase. This kind of devise is called chemostat.

2. *Turbidostat*: An electrical devise will monitor the cell density (turbidity) and at appropriate turbidity the nutrient will be added (as discussed above – glucose) at a dilution rate to maintain the constant population. If the cell density is higher, the instrument will add more quantity of solution to reduce or dilute the population and vice versa. This kind of device is referred as turbidostat.

Both the devices are highly useful in the continuous culture of bacteria.

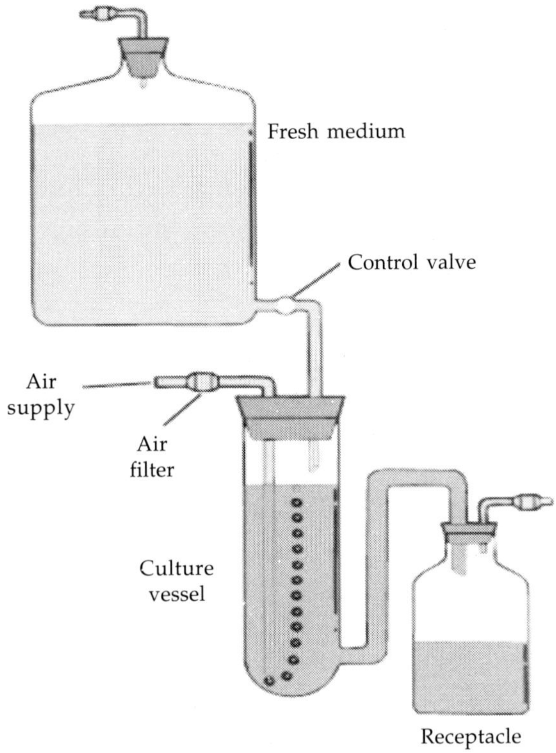

Chemostat – continuous culture system

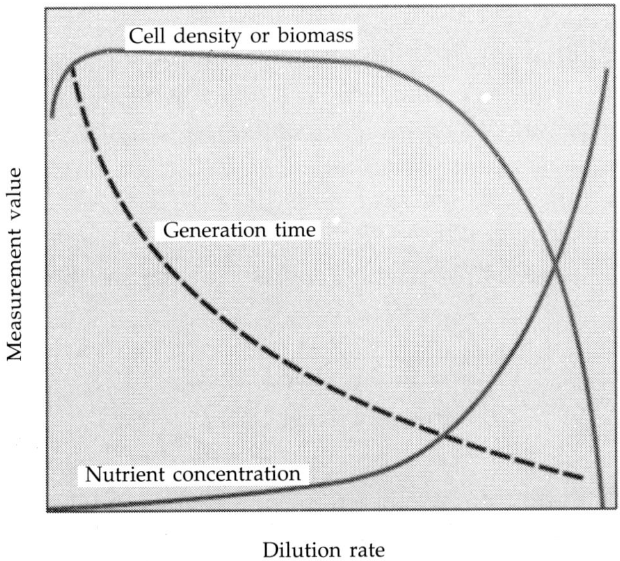

Chemostat – Dilution rate and microbial growth

Calculation of Growth rate and Generation time

Normal reproduction in bacteria is by binary fission

Increase in population is given by, $1 \rightarrow 2 \rightarrow 2^2 \rightarrow 2^3 \rightarrow 2^4 \rightarrow - - - - \rightarrow 2^n$

n = the number of generations.

Total population at the end of given period of time is N

$N = 1 \times 2^n$, If N_0 is the number of bacteria inoculated at time zero, (1)

$N = N_0 \times 2^n$ (2)

To solve the equation for number of generations (n)

$$\log_{10} N = \log_{10} N_0 + n \log_{10} 2$$

$$n = \frac{\log_{10} N - \log_{10} N_0}{\log_{10} 2} \qquad (3)$$

Substitute value of $\log_{10} 2$ by 0.301

$$n = \frac{\log_{10} N - \log_{10} N_0}{0.301}$$

$$n = 3.3 (\log_{10} N - \log_{10} N_0) \qquad (4)$$

Now n gives the number of generations

Generation time (g)

Time required for doubling of the cell number

Generation Time Determination. The generation time can be determined from a microbial growth curve. The population data are plotted with the logarithmic axis used for the number of cells. The time to double the population number is then read directly from the plot. The log of the population number can also be plotted against time on regular axes.

Using equation 4 for n, the generation time can be calculated as below,

$$g = \frac{t}{n} = \frac{t}{3.3 (\log_{10} N - \log_{10} N_0)}$$

where,

t- time interval

n-Number of generations

N- Bacterial count after t time

N_0- Bacterial count at initial count

For example: If time duration is 7 hours, log of initial count is 1.92 and log of final count is 8.53 the generation time is = 21.96 generations in 7 hours = 7 / 3.3(8.53 - 1.92) = *0.32 hours (x 60 = 19.2 minutes)*

Growth rate

It is the number of generations per hour. Increase of bacterial cell number per unit time / Number of generations per hour. It is the reciprocal of generation time g.

$$R = \frac{n}{t} = \frac{3.3\,(\log_{10} N - \log_{10} N_0)}{t}$$

Therefore in this case, 3.3(8.53 - 1.92) / 7= 3.12 generations / hr.

❏❏❏

Chapter 8

Nutritional Requirements for the Growth of Bacteria

Every organism must find in its environment all of the substances required for energy generation and cellular biosynthesis. The chemicals and elements of this environment that are utilized for bacterial growth are referred to as *nutrients* or *nutritional requirements*. Many bacteria can be grown the laboratory in *culture media* which are designed to provide all the essential nutrients in solution for bacterial growth.

Bacteria that are symbionts or obligate intracellular parasites of other cells, usually eucaryotic cells, are (not unexpectedly) difficult to grow outside of their natural host cells. Many bacteria can be identified in the environment by inspection or using genetic techniques, but attempts to isolate and grow them in artificial culture has been unsuccessful. This, in part, is the basis of the estimate that we may know less than one percent of all prokaryotes that exist.

The common nutrient requirements

Analysis of microbial cell composition shows that over 95% of cell dry weight is made up of a few major elements: carbon, oxygen, hydrogen, nitrogen, sulfur, phosphorus, potassium, calcium, magnesium, and iron. These are called *macroelements* or macronutrients because they are required by microorganisms in relatively large amounts. The first six (C, O, H, N, S, and P) are components of carbohydrates, lipids, proteins, and nucleic acids. The remaining four macroelements exist in the cell as cations and play a variety of roles. For example, potassium (K^+) is required for activity by a number of enzymes, including some of those involved in protein synthesis. Calcium (Ca^{2+}), among other functions, contributes to the heat resistance of bacterial endospores. Magnesium (Mg^{2+}) serves as a cofactor for many enzymes, complexes with ATP, and stabilizes ribosomes and cell membranes. Iron (Fe^{2+} and Fe^{3+}) is a part of cytochromes and a cofactor for enzymes and electron-carrying proteins.

All organisms, including microorganisms, require several *micronutrients* or *trace elements* besides macroelements. The micronutrients - manganese, zinc, cobalt, molybdenum, nickel, and copper - are needed by most cells. However, cells require such small amounts that contaminants in water, glassware, and regular media components often are adequate for growth. Therefore it is very difficult to demonstrate a micronutrient requirement.

In nature, micronutrients are ubiquitous and probably do not usually limit growth. Micronutrients are normally a part of enzymes and cofactors, and they aid in the catalysis of reactions and maintenance of protein structure. For example, zinc (Zn^{2+}) is present at the active site of some enzymes but is also involved in the association of regulatory and catalytic subunits in *E. coli* aspartate carbamoyltransferase. Manganese (Mn^{2+}) aids many enzymes catalyzing the transfer of phosphate groups. Molybdenum (Mo^{2+}) is required for nitrogen fixation, and cobalt (Co^{2+}) is a component of vitamin B12.

Besides the common macro elements and trace elements, microorganisms may have particular requirements that reflect the special nature of their morphology or environment. If an essential nutrient is in short supply, microbial growth will be limited regardless of the concentrations of other nutrients.

Six basic nutrients for bacterial growth

The majority of living cells, especially bacteria, need over 20 chemical elements in order to function. Six elements in particular are critical for building biological macromolecules and maintaining their structure and function.

All life has the same basic nutritional requirements, which include:

1. *A source of energy*. This may be light (the sun or lamps) or inorganic substances like sulfur, carbon monoxide or ammonia, or preformed organic matter like sugar, protein, fats etc. Without energy life cannot exist and quickly dies or becomes inactive.
2. *Hydrogen* occurs in organic hydrocarbon molecules and in inorganic molecules such as water. A major role of hydrogen in biological systems is to maintain structure and conformation of macromolecules. Hydrogen bonds contribute to base-pairing between complementary DNA strands and to folding of proteins.
3. *A source of nitrogen*. This may be nitrogen gas, ammonia, nitrate/ nitrite, or a nitrogenous organic compound like protein or nucleic acid. Nitrogen is used in protein/ amino acid synthesis and nucleic acid synthesis. Nitrogen-containing bases help spell out information contained within the genetic code. A nitrogenous base (adenine) is a building block of the energy-rich molecule, ATP. Peptidoglycan is also a nitrogen-rich substance.
4. *A source of carbon*. This can be carbon dioxide or monoxide, methane, carbon monoxide, or complex organic material. Carbon is needed to build all organic compounds. All organisms require a carbon source for this purpose and many microbes also use some form of carbon as their energy source.
5. *A source of oxygen*. All cells use oxygen in a bound form and many require gaseous oxygen (air), but oxygen is lethal to many microbes.
6. *A source of phosphorous, sulfur, magnesium, potassium & sodium.* Phosphorus is essential for nucleic acid synthesis and formation of phospholipids that contribute to the integrity of cell membranes. Sulfur occurs in certain amino acids such as cysteine and methionine. Disulfide bonds help maintain tertiary and quaternary levels of protein structure and are crucial in maintaining the function of many proteins. A source of minerals like iron, zinc, cobalt etc. These are called TRACE metals that are required by some enzymes to function.
7. *A source of calcium.* Most cells require calcium in significant quantities, but some seem to only need it in trace amounts.
8. *A source of water.* All life requires liquid water in order to grow and reproduce; Some resting stages of cells, like bacterial spores,

can exist for long periods without free water, but they do not grow or metabolize.

The sources of these various requirements *define an organism*, so a description of every organism should include this information. Many bacteria can synthesize every complex molecule they need from the *basic minerals*, but others, said to be *fastidious*, require *preformed* organic molecules like vitamins, amino acids, nucleic acids, carbohydrates

Overview of the microbial nutritional requirements

Major and trace elements, their sources and functions in bacterial cells

Element	% of dry weight	Source	Some Laboratory reagents used in media preparation	Function
Major elements				
Carbon	50	organic compounds or CO_2	Glucose, sucrose, lactose, malic acid, mannitol, starch, dextrin,	Main constituent of cellular material
Oxygen	20	H_2O, organic compounds, CO_2, and O_2	Water, air and organic compounds	Constituent of cell material and cell water; O_2 is electron acceptor in aerobic respiration
Hydrogen	8	H_2O, organic compounds, H_2	Water, air and organic compounds	Main constituent of organic compounds and cell water
Nitrogen	14	NH_3, NO_3, organic compounds, N_2	Ammonium sulphate, sodium sulphate, sodium nitrate, sodium nitrite, peptone, urea, tryptone, soybean extract, beef extract, yeast extract	Constituent of amino acids, nucleic acids nucleotides, and coenzymes
Phosphorus	3	inorganic phosphates (PO_4)	KH_2PO_4, K_2HPO_4, NaH_2PO_4, Na_2HPO_4	Constituent of nucleic acids, nucleotides, phospholipids, LPS, teichoic acids

Sulfur	1	SO_4, H_2S, So, organic sulfur compounds	Na_2 Na_2 ami
Potassium	1	Potassium salts	KCl K_2F
Magnesium	0.5	Magnesium salts	Mg
Calcium	0.5	Calcium salts	CaC
Iron	0.2	Iron salts	FeS FeC

Apart from these macro and micro nutrients, the microorganisms may require a small amount of certain organic compounds for growth, because of their essential and unable to synthesis from available nutrients.

Common vitamins required in the nutrition of certain bacteria

Vitamin	Coenzyme form
p-Aminobenzoic acid (PABA)	-
Folic acid	Tetrahydrofolate
Biotin	Biotin
Lipoic acid	Lipoamide
Mercaptoethane-sulfonic acid	Coenzyme M
Nicotinic acid	NAD (nicotinamide aden dinucleotide) and NADP
Pantothenic acid	Coenzyme A and the Acy Carrier Protein (ACP)
Pyridoxine (B6)	Pyridoxal phosphate

Chapter 9

Conditions for Growth of Bacteria

As we have seen previously, microorganisms must be able to respond to variations in nutrient levels, and particularly to nutrient limitation. The growth of microorganisms also is greatly affected by the chemical and physical nature of their surroundings. An understanding of environmental influences aids in the control of microbial growth and the study of the ecological distribution of microorganisms.

The ability of some microorganisms to adapt to extreme and inhospitable environments is truly remarkable. Prokaryotes are present anywhere life can exist. Many habitats in which prokaryotes thrive would kill most other organisms. Prokaryotes such as *Bacillus infernus* even seem able to live over 1.5 miles below the Earth's surface, without oxygen and at temperatures above 60°C.

Microorganisms that grow in such harsh conditions are often called *extremophiles*.

In this section we shall briefly review how some of the most important environmental factors affect microbial growth. Major emphasis will be given to range of physical conditions such as O_2 concentration, hydrogen ion concentration (pH), temperature, salt concentration and water activity.

A. Temperature

Microorganisms have been found growing in virtually all environments where there is liquid water, regardless of its temperature. In 1966, Professor Thomas D. Brock at Indiana University made the amazing discovery in *boiling hot springs* of *Yellowstone National Park* that bacteria were not just surviving there, they were growing and flourishing. Boiling temperature could not inactivate any essential enzyme. Subsequently, prokaryotes have been detected growing around black smokers and hydrothermal vents in the deep sea at temperatures at least as high as 115°C. Microorganisms have been found growing at very low temperatures as well. In super cooled solutions of H_2O as low as -20°C, certain organisms can extract water for growth, and many forms of life flourish in the icy waters of the Antarctic, as well as household refrigerators, near 0°C.

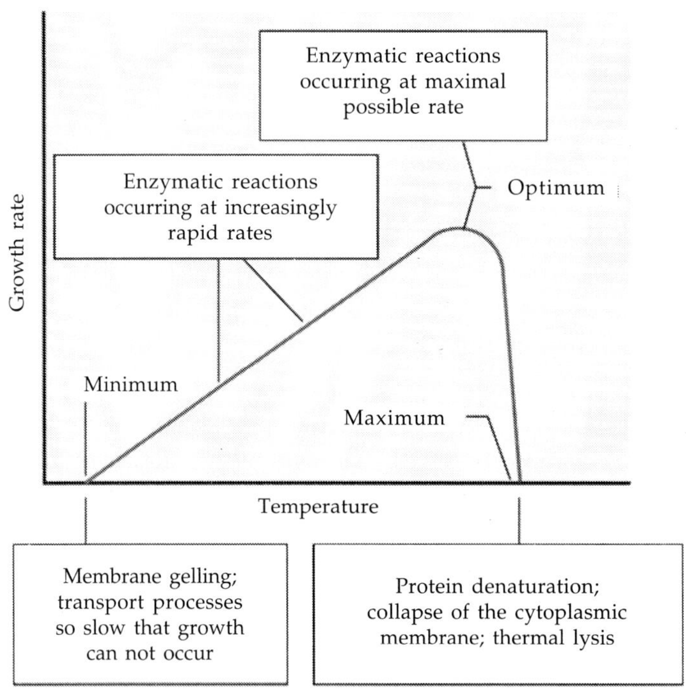

Bacterial growth at different temperature conditions

Temperature classes of bacteria

Microorganisms can be placed in one of five classes based on their temperature ranges for growth.

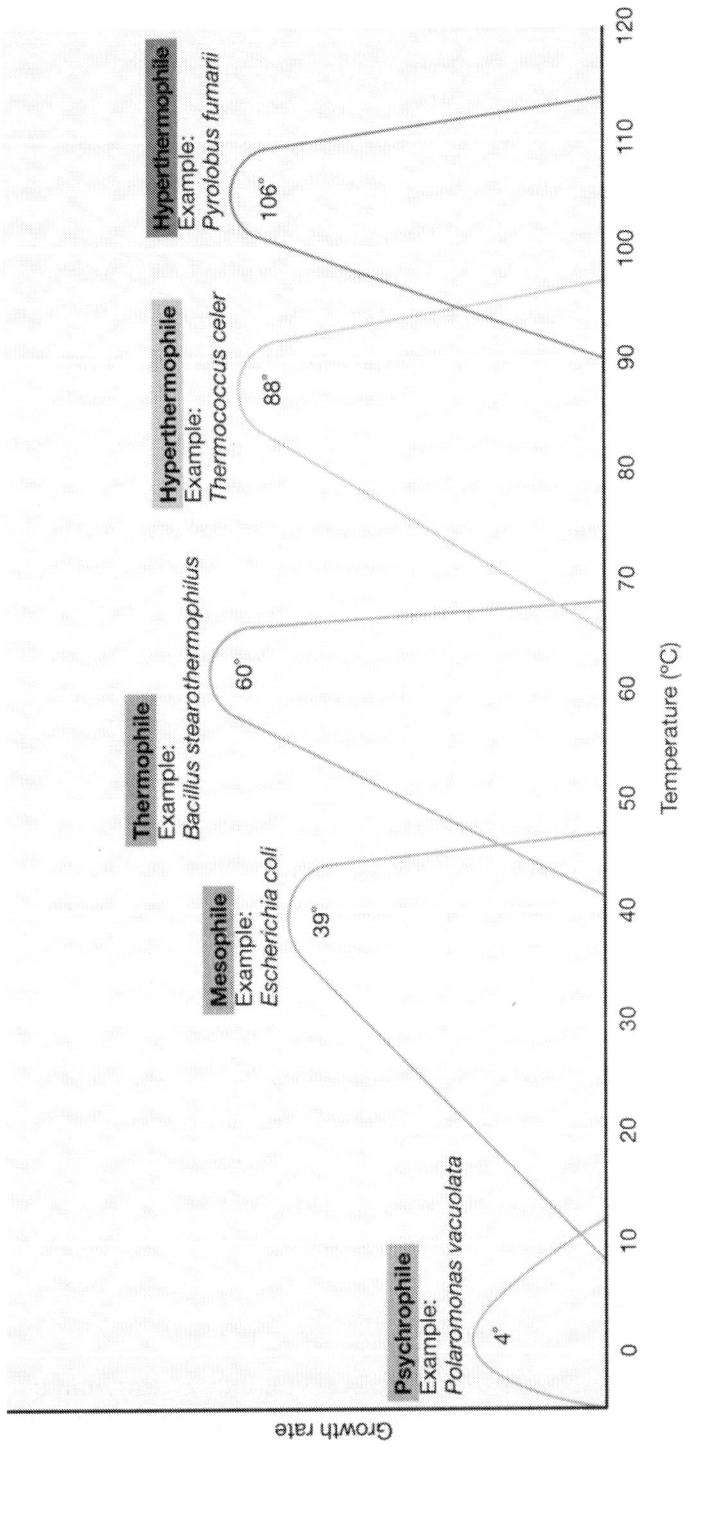

Temperature ranges for microbial growth

Min, max and opt. temperature requirements of some of the organism

Bacterium	Minimum (°C)	Optimum (°C)	Maximum (°C)
Listeria monocytogenes	1	30-37	45
Vibrio marinus	4	15	30
Pseudomonas maltophilia	4	35	41
Thiobacillus novellus	5	25-30	42
Staphylococcus aureus	10	30-37	45
Escherichia coli	10	37	45
Clostridium kluyveri	19	35	37
Streptococcus pyogenes	20	37	40
Streptococcus pneumoniae	25	37	42
Bacillus flavothermus	30	60	72
Thermus aquaticus	40	70-72	79
Methanococcus jannaschii	60	85	90
Sulfolobus acidocaldarius	70	75-85	90
Pyrobacterium brockii	80	102-105	115

1. Psychrophiles

They grow well at 0°C and have an optimum growth temperature of 15°C or lower; the maximum is around 20°C. They are readily isolated from Arctic and Antarctic habitats; because 90% of the ocean is 5°C or colder, it constitutes an enormous habitat for psychrophiles. The psychrophilic alga *Chlamydomonas nivalis* can actually turn a snowfield or glacier pink with its bright red spores. Psychrophiles are widespread among bacterial taxa and found in such genera as *Pseudomonas, Vibrio, Alcaligenes, Bacillus, Arthrobacter, Moritella, Photobacterium,* and *Shewanella*. The psychrophilic archaeon *Methanogenium* has recently been isolated from Ace Lake in Antarctica. Psychrophilic microorganisms have adapted to their environment in several ways. Their enzymes, transport systems, and protein synthetic mechanisms function well at low temperatures. The cell membranes of psychrophilic microorganisms have high levels of unsaturated fatty acids and remain semifluid when cold. Indeed, many psychrophiles begin to leak cellular constituents at temperatures higher than 20°C because of cell membrane disruption.

2. Psychrotrophs

Many species can grow at 0 to 7°C even though they have optima between 20 and 30°C, and maxima at about 35°C. These are called *psychrotrophs* or *facultative psychrophiles*. Psychrotrophic bacteria and fungi are major factors in the spoilage of refrigerated foods.

3. Mesophiles

Mesophiles are microorganisms with growth optima around 20 to 45°C; they often have a temperature minimum of 15 to 20°C. Their maximum is about 45°C or lower. Most microorganisms probably fall within this category. Almost all human pathogens are mesophiles, as might be expected since their environment is a fairly constant 37°C.

4. Thermophiles

Some microorganisms are thermophiles; they can grow at temperatures of 55°C or higher. Their growth minimum is usually around 45°C and they often have optima between 55 and 65°C. The vast majority are prokaryotes although a few algae and fungi are thermophilic. These organisms flourish in many habitats including composts, self-heating hay stacks, hot water lines, and hot springs. Thermophiles differ from mesophiles in having much more heat-stable enzymes and protein synthesis systems able to function at high temperatures. Their membrane lipids are also more saturated than those of mesophiles and have higher melting points; therefore thermophile membranes remain intact at higher temperatures.

5. Hyperthermophiles

As mentioned previously, a few thermophiles can grow at 90°C or above and some have maxima above 100°C. Prokaryotes that have growth optima between 80°C and about 113°C are called hyperthermophiles. They usually do not grow well below 55°C. *Pyrococcus abyssi* and *Pyrodictium occultum* are examples of marine hyperthermophiles found in hot areas of the seafloor.

B. pH

pH is a measure of the hydrogen ion activity of a solution and is defined as the negative logarithm of the hydrogen ion concentration (expressed in terms of molarity).

$$pH = -\log [H^+] = \log(1/[H^+])$$

The pH scale extends from pH 0.0 to pH 14.0, and each pH unit represents a tenfold change in hydrogen ion concentration. It is not surprising that pH dramatically affects microbial growth. Each species has a definite pH growth range and pH growth optimum.

1. Acidophiles have their growth optimum between pH 0 and 5.5
2. Neutrophiles, between pH 5.5 and 8.0
3. Alkalophiles prefer the pH range of 8.5 to 11.5.
4. Extreme alkalophiles have growth optima at pH 10 or higher

In general, different microbial groups have characteristic pH preferences. Most bacteria and protozoa are neutrophiles. Most fungi prefer slightly acid surroundings, about pH 4 to 6; algae also seem to favor slight acidity. There are many exceptions to these generalizations. For example, the alga *Cyanidium caldarium* and the archaeon *Sulfolobus acidocaldarius* are common inhabitants of acidic hot springs; both grow well around pH 1 to 3 and at high temperatures. The Archaea *Ferroplasma acidarmanus* and *Picrophilus oshimae* can actually grow at pH 0, or very close to it.

pH	$[H^+]$ (Molarity)		Environmental examples	Microbial examples
0	10^{-0}(1.0)	Increasing acidity ↑	Concentrated nitric acid	*Ferroplasma* *Picrophilus oshimae*
1	10^{-1}		Gastric contents, acid thermal springs	*Dunaliella acidophila*
2	10^{-2}		Lemon juice Acid mine drainage	*Cyanidium caldarium* *Thiobacillus thiooxidans* *Suffiolobus acidocaldarius*
3	10^{-3}		Vinegar, ginger ale Pineapple	
4	10^{-4}		Tomatoes, orange juice Very acid soil	
5	10^{-5}		Cheese, cabbage Bread	*Physarum polycephalum* *Acanthamoeba castellanii*
6	10^{-6}		Beef, chicken Rain water Milk Saliva	*Lactobacillus acidophilus* *E. coli, Pseudomonas aeruginosa,* *Euglena gracilis, Paramecium bursaria*
7	10^{-7}	Neutrality	Pure water Blood	*Staphyloccus aureus*
8	10^{-8}		Seawater	*Nitrosomonas* spp.
9	10^{-9}		Strongly alkaline soil Alkaline lakes	
10	10^{-10}		Soap	*Microcystis aeruginosa* *Bacillus alcalophilus*
11	10^{-11}		Household ammonia	
12	10^{-12}		Saturated calcium hydroxide solution	
13	10^{-13}		Bleach Drain opener	
14	10^{-14}	↓ Increasing alkalinity		

The pH scale

C. Oxygen

Oxygen is a universal component of cells and is always provided in large amounts by H_2O. However, prokaryotes display a wide range of responses to molecular oxygen O_2. Molecular oxygen is both beneficial and harmful for biologicals.

Beneficial: It act as terminal electron acceptor in the respiratory chain – Electron transport chain reaction enable to form ATP.

Harmful: The oxygen derivatives such as hydrogen *peroxide, singlet oxygen and super oxide* radical are highly toxic and oxidizing agents.

Oxygen classes of bacteria

The organisms can be divided into 4 or 5 different groups based on their oxygen requirement.

1. Obligate aerobes

They require O_2 for growth; they use O_2 as a final electron acceptor in aerobic respiration. Ex. *Micrococcus*

2. Obligate anaerobes

Obligate anaerobes do not need or use O_2 as a nutrient. In fact, O_2 is a toxic substance, which either kills or inhibits their growth. Obligate anaerobic prokaryotes may live by fermentation, anaerobic respiration, bacterial photosynthesis, or the novel process of methanogenesis. Ex. *Clostridium, Methanobcaterium*

3. Facultative anaerobes (or facultative aerobes)

They are organisms that can switch between aerobic and anaerobic types of metabolism. Under anaerobic conditions (no O_2) they grow by fermentation or anaerobic respiration, but in the presence of O_2 they switch to aerobic respiration. Ex. *E. coli*

4. Aerotolerant anaerobes

They are bacteria with an exclusively anaerobic (fermentative) type of metabolism but they are insensitive to the presence of O_2. They live by fermentation alone whether or not O_2 is present in their environment. Ex. *Streptococcus*

5. Microaerophiles

They need very low level of oxygen for their growth, but cannot tolerate the level of atmospheric O_2 level. Ex. *Spirillum volutans*

Terms used to describe O_2 Relations of Microorganisms

	Environment		
Group	**Aerobic**	**Anaerobic**	**O_2 Effect**
Obligate Aerobe	Growth	No growth	Required (utilized for aerobic respiration)
Microaerophile	Growth if level not too high	No growth	Required but at levels below 0.2 atm
Obligate Anaerobe	No growth	Growth	Toxic
Facultative Anaerobe (Facultative Aerobe)	Growth	Growth	Not required for growth but utilized when available
Aerotolerant Anaerobe	Growth	Growth	Not required and not utilized

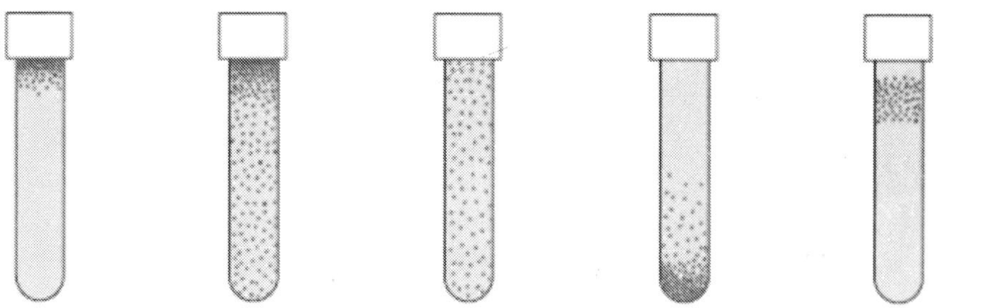

Bacteria and oxygen concentration

Oxygen detoxification

Most of the aerobes (and aerotolerant organisms too) have three major enzymes either all or two to detoxify the oxygen derivatives. They are **catalase, peroxidase and super oxide dismutase.** Obligate anaerobes lack these enzymes, and therefore undergo lethal oxidations by various oxygen radicals when they are exposed to O_2.

D. Water

All living cells, including bacteria, require an aqueous environment. However certain structures formed by microbes are environmentally resistant and can withstand desiccation for long periods. These include bacterial endospores. Osmosis is the tendency for water to move in and out of the cell to equalize it- cytoplasm more concentrated than pure water so water wants to move in if cell is suspended in pure water. A hypertonic environment has a high solute concentration with respect to the inside of a cell.

Consequently there is more water inside the cell with respect to the exterior and water tends to diffuse out of the cell. The result is the cell shrinks because of plasmolysis. Salting can be used to preserve foods and takes advantage of this important principle. The higher the water content of a food the more it is said to be "potentially hazardous." That is the high moisture content favors bacterial growth and is the food is prone to spoilage. Archaebacteria known as halophiles are salt-loving bacteria found in oceans and salt marshes that actively pump salt into their cells. Halophiles require at least 15% salt concentration (0.9%is intracellular concentration of most cells), prefer 25% and will tolerate a 30% concentration. Examples: Great Salt Lake, pickle spoilers.

Functions of water

- Water serves as medium in which food for the organisms is dissolved for entering into the cell for metabolism.
- Carries away waste products of metabolism from the cell.
- Water is essential in metabolic reactions such as the hydrolysis of proteins, fats etc.
- It is the final end product of Oxidation of hydrogen from organic compounds to yield energy.

Water activity is a measure of unbound or free water available for biological and chemical reactions. Water activity is defined as the ratio of partial vapour pressure of water in food (P) to the vapour pressure of pure water (Po) at a given temperature.

The term is designated by Scot (1953). aw = ERH/100. Relative humidity is the ratio of partial vapour pressure of water in air to vapour pressure of pure water at a given temperature.

- Most bacteria do not grow below 0.90 aw,
- Most yeasts do not grow below 0.80 aw,
- Most molds do not grow below 0.70 aw,
- No growth of microbes and no spoilage of food below 0.65-0.70 aw.

aw	Importance
0.90	Lower limit for bacterial growth
0.85	Many yeasts inhibited
0.80	Lower limit for the growth of yeasts
0.75	Lower limit for halophilic bacteria
0.70	Lower limit for xerophilic fungi
0.65	Maximum velocity of Maillard reaction
0.60	Lower limit for osmophilic and xerophilic yeasts and molds
0.55	Lower limit for life
0.40	Minimum oxidative velocity

Minimum water activity for food borne pathogens

Bacteria	Min. water activity for growth
E.coli	0.950
Staphylococcus aureus	0.860
Salmonella sp.	0.950
Bacillus subtilis	0.950
Pseudomonas fluorescens	0.957

Microorganisms differ greatly in their ability to adapt to habitats with low water activity. A microorganism must expend extra effort to grow in a habitat with a low aw value because it must maintain a high internal solute concentration to retain water. Some microorganisms can do this and are *osmotolerant*; they will grow over wide ranges of water activity or osmotic concentration. For example, *Staphylococcus aureus* can be cultured in media containing any sodium chloride concentration up to about 3 M. It is well adapted for growth on the skin. The yeast *Saccharomyces rouxii* will grow in sugar solutions with aw values as low as 0.6. The alga *Dunaliella viridis* tolerates sodium chloride concentrations from 1.7 M to a saturated solution. Many fungi are osmotolerant and thus particularly important in the spoilage of salted or dried foods.

Xerophiles are organisms which live in dry environments (made dry by lack of water).

E. Salt Concentrations

The only common solute in nature that occurs over a wide concentration range is salt [NaCl], and microorganisms are named based on their growth response to salt. Microorganisms that require some NaCl for growth are *halophiles*.

Halophiles have adapted so completely to hypertonic, saline conditions that they require high levels of sodium chloride to grow, concentrations between about 2.8 M and saturation (about 6.2 M) for extreme halophilic bacteria. The archaeon *Halobacterium* can be isolated from the Dead Sea (a salt lake between Israel and Jordan and the lowest lake in the world), the Great Salt Lake in Utah, and other aquatic habitats with salt concentrations approaching saturation. *Halobacterium* and other extremely halophilic bacteria have significantly modified the structure of their proteins and membranes rather than simply increasing the intracellular concentrations of solutes, the approach used by most osmotolerant microorganisms. These extreme halophiles accumulate enormous quantities of potassium in order to remain hypertonic to their environment; the internal potassium concentration may reach 4 to 7 M. The enzymes, ribosomes, and transport proteins of these bacteria require high levels of potassium for stability and activity. In addition, the plasma membrane and cell wall of *Halobacterium* are stabilized by high concentrations of sodium ion. If the sodium concentration decreases too much, the wall and plasma membrane literally disintegrate. Extreme halophilic bacteria have successfully adapted to environmental conditions that would destroy most organisms. In the process they have become so specialized that they have lost ecological flexibility and can prosper only in a few extreme habitats.

1. Mild halophiles require 1-6% salt, (Ex: *Staphylococcus areus*)
2. Moderate halophiles require 6-15% salt; (Ex. *Vibrio fisheri*
3. Extreme halophiles require 15-30% NaCl for growth. (Ex. *Halobacterium salanarum*)

Bacteria that are able to grow at moderate salt concentrations, even though they grow best in the absence of NaCl, are called *halotolerant*. Although halophiles are "osmophiles" (and halotolerant organisms are "osmotolerant") the term *osmophiles* is usually reserved for organisms that are able to live in environments high in sugar.

F. Pressure

Most organisms spend their lives on land or on the surface of water, always subjected to a pressure of 1 atmosphere (atm), and are never affected significantly by pressure. Yet the deep sea (ocean of 1,000 m or more in depth) is 75% of the total ocean volume. The hydrostatic pressure can reach 600 to 1,100 atm in the deep sea, while the temperature is about 2 to 3°C.

Despite these extremes, bacteria survive and adapt. Many are,

a. *Barotolerant* - increased pressure does adversely affect them but not as much as it does nontolerant bacteria. Some bacteria in the gut of deep-sea invertebrates such as amphipods and holothurians.

b. *Barophilic* - they grow more rapidly at high pressures. These gut bacteria may play an important role in nutrient recycling in the deep sea. One barophile has been recovered from the Mariana trench near the Philippines (depth about 10,500 m) that is actually unable to grow at pressures below about 400 to 500 atm when incubated at 2°C.Thus far, barophiles have been found among several bacterial genera (e.g., *Photobacterium, Shewanella, Colwellia*). Some members of the Archaea are thermobarophiles (e.g., *Pyrococcus* spp., *Methanococcus jannaschii*).

An overview on microbial responses to environmental factors

Descriptive term	Definition	Representative microorganisms
Solute and Water Activity		
Osmotolerant	Able to grow over wide ranges of water activity or osmotic	*Staphylococcus aureus, Saccharomyces routii*
Halophile	Requires high levels of sodium chloride, usually above about 0.2 M, to grow	*Halobacterium, Dunaliella, Ectothiorhodospira*
pH		
Acidophile	Growth optimum between pH 0 and 5.5	*Suifolobus, Picrophilus, Ferroplasma, Acontium, Cyanidium caldanium*
Neutrophile	Growth optimum between pH 5.5 and 8.0	*Ercherichia, Eaglena, Paramecium*
Alkalophile	Growth optimum between pH 8.5 and 11.5	*Bacillus alcalophilus, Natronobacterium*
Temperature		
Psychrophile	Grows well at 0°C and has an optimum growth temperature of 15°C or lower	*Bacillus psychrophilus, Chlamydomonas nivalis*
Psychrotroph	Can grow at 0-7°C; has an optimum between 20 and 30°C and a maximum around 35°C	*Listeria monocytogenes, Pseadomonas fluorescens*
Mesophile	Has growth optimum around 20-45°C	*Escherichia coli, Neisseria gonorrhoeae, Trichomonas vaginalis*
Themophile	Can grow at 55°C or higher; optimum often between 55 and 65°C	*Bacillus stearothermophilus, Thermus aquaticus, Cyanidium coldarium, Chaetomium thermophile*
Hyperthermophile	Has an optimum between 80 and about 113°C	*Sulfolobus, Pyrococcus, Pyrodictium*
Oxygen Concentration		
Obligate aerobe	Completely dependent on atmospheric O_2 for growth	*Micrococus lateus, Pseudomonas, Mycobacterium;* most algae, fungi and protozoa
Faculative anaerobe	Does not require O_2 for growth, but grows better is its presence	*Escherichia, Enterococcus, Saccharomyces cervisiae*
Aerotolerant anaerobe	Grows equally well in presence or absence of O_2	*Streptococcus pyogenes*
Obligate anaerobe	Does not tolerate O_2 and dies in its presence	*Clostridium, Bacteroides, Methamobacterium, Trepomonas agilis*
Microaerophile	Requires O2 levels below 2-10% for growth and is damaged by atmospheric O_2 (20%)	*Campylobacter, Spirillum volutons, Treponema pallidum*
Pressure		
Barophilic	Growth more rapid at high hydrostatic pressure	*Photobacterium profundum, Shewanella benthica*

❑❑❑

Chapter 10

Microbial Control

Although many microorganisms are beneficial and necessary for human well-being, microbial activities may have undesirable consequences, such as food spoilage and disease. Therefore it is essential to be able to kill a wide variety of microorganisms or inhibit their growth to minimize their destructive effects. The goal is twofold: (1) to destroy pathogens and prevent their transmission, and (2) to reduce or eliminate microorganisms responsible for the contamination of water, food, and other substances. This chapter focuses on the control of microorganisms by nonspecific physical and chemical agents.

Control

Reduction in numbers and or activity of the total microbial flora. Control can be achieved by inhibition, removal or killing with physical agents, physical process or chemical agents. *Antimicrobial agent* is the one that interferes with growth and metabolism of the microorganisms include antibacterial and antifungal agents. Antimicrobial agents used to treat infections are called *chemotherapeutic agents.*

Reasons for control

- To prevent transmission of disease and infection
- To prevent contamination by or growth of undesirable micro-organsms
- To prevent deterioration and spoilage of materials

Mode of action of antimicrobial agents

- Damage to the cell wall or inhibition of cell wall synthesis
- Alteration of the physical or chemical state of proteins and nucleic acids
- Alteration of the permeability of the cytoplasmic membrane
- Inhibition of enzyme action
- Inhibition of protein or nucleic acid synthesis

Conditions influencing antimicrobial action

1. Environment: Has got profound influence on the rate as well as efficacy of the microbial destruction e.g. Effectiveness of heat is much greater in acid than in alkaline material
2. Kinds of microorganisms: Species differ in their susceptibility to physical and chemical agents. Vegetative cells are more susceptible than sporulated cells
3. Physiological state of the cell: young, actively metabolizing cells are easily destroyed than old dormant cells. Non growing cells would not be affected.

Terminologies used mostly in microbial control

Terminology is especially important when the control of microorganisms is discussed because words like disinfectant and antiseptic often are used loosely. The situation is even more confusing because a particular treatment can either inhibit growth or kill depending on the conditions.

The ability to control microbial populations on inanimate objects, like eating utensils and surgical instruments, is of considerable practical importance. Sometimes it is necessary to eliminate all microorganisms from an object, whereas only partial destruction of the microbial population may be required in other situations.

1. *Sterilization* [Latin *sterilis,* unable to produce offspring or barren] is the process by which all living cells, viable spores, viruses and viroids are either destroyed or removed from an object or habitat. A sterile object is totally free of viable microorganisms, spores, and other infectious agents. When sterilization is achieved by a chemical agent, the chemical is called a sterilant.

2. *Disinfection* is the killing, inhibition, or removal of microorganisms that may cause disease. The primary goal is to destroy potential pathogens, but disinfection also substantially reduces the total microbial population.

3. *Disinfectants* are agents, usually chemical, used to carry out disinfection and are normally used only on inanimate objects. A disinfectant does not necessarily sterilize an object because viable spores and a few microorganisms may remain.

4. *Sanitization* is closely related to disinfection. In sanitization, the microbial population is reduced to levels that are considered safe by public health standards. The inanimate object is usually cleaned as well as partially disinfected. For example, sanitizers are used to clean eating utensils in restaurants. It is frequently necessary to control microorganisms on living tissue with chemical agents.

5. *Antisepsis* [Greek *anti,* against, and *sepsis,* putrefaction] is the prevention of infection or sepsis and is accomplished with *antiseptics*. These are chemical agents applied to tissue to prevent infection by killing or inhibiting pathogen growth; they also reduce the total microbial population. Because they must not destroy too much host tissue, antiseptics are generally not as toxic as disinfectants.

6. *Cide* - A suffix can be employed to denote the type of antimicrobial agent. Substances that kill organisms often have the suffix –cide [Latin *cida,* to kill]: *Germicide* kills pathogens (and many nonpathogens) but not necessarily endospores. A disinfectant or antiseptic can be particularly effective against a specific group, in which case it may be called a *bactericide, fungicide, algicide,* or *viricide*. Other chemicals do not kill, but they do prevent growth.

7. *Static* - If these agents are removed, growth will resume. Their names end in -static [Greek *statikos,* causing to stand or stopping]—for example, *bacteriostatic* and *fungistatic*.

Although these agents have been described in terms of their effects on pathogens, it should be noted that they also kill or inhibit the growth of nonpathogens as well. Their ability to reduce the total microbial population, not just to affect pathogen levels, is quite important in many situations.

Physical agents for microbial control

Heat and other physical agents are normally used to control microbial growth and sterilize objects, as can be seen from the continual operation of the autoclave in every microbiology laboratory. The four most frequently employed physical agents are heat, low temperatures, filtration and radiation.

1. Heat / high temperature

Fire and boiling water have been used for sterilization and disinfection since the time of the Greeks, and heating is still one of the most popular ways to destroy microorganisms. Either moist or dry heat may be applied.

The heat susceptibility of the organism should be known before applying the heat treatment. For this, the thermal death time and thermal death points are essential.

Thermal death time refers the time to kill a suspension of organisms at a specific temperature.

Thermal death point refers the lowest temperature to kill a suspension in a given time.

Decimal reduction time (D time) – Time to kill 90% or 1 log unit of a population at a given temperature

Based on the above parameters, the time and temperature to kill the microorganism from an environment is derived. Practically heat can be applied either dry or moist heat for sterilization.

A. Moist heat

(i) Steam under pressure (Moist heat)

Heat in the form of saturated steam under pressure is the most practical and dependable agent for sterilization. Steam under pressure provides temperature above boiling. In addition, it has the advantage of rapid heating, penetration and moisture in abundance which facilitates coagulation of proteins.Moist heat causes denaturation and coagulation of protein. To kill vegetative structure of bacteria, yeast and molds 80°C for 5-10 min; To kill mold spores 80°C for 30 min; bacterial spores 121°C, 15 lbs/sq.inch pressure for 15 min. are the optimum moist heat.

(The moist heat has more penetration power than dry heat and leads faster reduction in number of living organism and under pressure, the penetration of heat will be more in the autoclave)

Instrument: The autoclave is used for sterilization using moist heat under pressure. Autoclave is made up of double walled steel plates with air tight lids. Electrical coil submerged in water is provided to generate the steam. A pressure gauge to measure the pressure, and a safety valve for safety, are also attached in the lid. If the steam pressure inside the closed vessel is increased to 15 lb/sq. inch, the temperature will raise to 121.6°C. Keeping this condition for 15-20 min. will kill all the vegetative and spore structures of microbes. Can also be done at 15 lbs (115C) for 30 minutes or at 20 lbs (126.6°C) for 10 minutes

Autoclave

(Note : For soil like heavy solid material, the penetration of moist heat is not sufficient to kill all the spores and vegetative structures. So intermittent sterilization (tyndallization) is necessary)

(ii) Boiling

Vegetative forms are killed in minutes. However, it is unreliable for killing spores.

(iii) Steam at atmospheric pressure

Steam has latent heat and at 100° C has 540 calories. Latent heat is released when steam condenses on a cold surface causing proteins to coagulate.

(iv) Pasteurization

It refers to removal of undesired microorganisms by heating at particular temperature without affecting the beneficial micro-organisms, odour and taste. It is a rapid method. The pasteurization of milk is done to eliminate the pathogenic microorganisms which cause diseases like tuberculosis, brucellosis, Q fever, typhoid etc. In broad, the pasteurization is done to kill *Salmonella* and *E. coli* like organisms.

- *Flash pasteurization*: holding the substance at 71°C for 15 seconds and rapid cooling referred as flash pasteurization. Also referred as High Temperature Short Time (HTST) method
- *Bulk pasteurization*: holding the substance at 63-65°C for 30 min and slow cooling referred as bulk pasteurization. Also referred as Low Temperature Holding (LTH) method.

(v) Tyndallization

Refers to intermittent or fractional sterilization. The subsequent cooling and heating by steam at 100°C for 3 days will remove the germs and their spores. Ex. Soil which contains diversified microbes with spores. On the first day all the vegetative cells will be killed and by the subsequent days germinated spores will be killed. The normal sterilization by autoclave will eliminate the only vegetative cells and not spores. The apparatus used for tyndallization is the steam Arnold, in which free flowing steam is used

Tyndallization > *sterilization* (by autoclave) > *pasteurization* is the order of degree of strength in terms of removal of microorganisms.

B. Dry heat

(i) Hot air sterilization

Causes oxidation of cells. Penetrating ability is less than moist heat. By using electrical coils, heat is generated and used for sterilization. Normally high temperature and long time are required for complete sterilization using this method. 160-180°C for 2- 3 hours are required for complete sterilization.

Instrument: Hot air oven. A double walled (inner and outer walls) instrument protected with asbestos sheets to avoid heat loss. Thermostat is present to control the temperature

Hot air oven

and timer is also present to turn off automatically after particular time. The glasswares like conical flasks, beakers, Petri dishes, pipettes, oils and powders and similar substances can be sterilized using this technique.

(ii) Direct flaming

Showing the objects or equipments directly to the flame leads the sterilization. This is also called as incineration.The inoculation needles, spread rods, forceps can be sterilized using this technique.

Advantages of heat sterilization

1. Sterilization is very effective.
2. Heat deliver system can be monitored effectively with various controls like pressure gauge, temperature meters etc.
3. Established quality control methods are available.

Disadvantages

1. Steam impermeable materials like fats, oils and powders can not be sterilized by autoclaving.
2. Heat sensitive materials like Serum, Antibiotics, Plastic materials, Vaccines and Rubbers can not be sterilized.
3. Presence of organic matters interfere with effective sterilization.
4. Dangers of explosion can not be sterilized by heat.

2. Low Temperature

Temperature below the optimum level always inhibit the growth and metabolism of microorganisms. Low temperature is used mainly as a mode for microbial preservation. This is only microbisatic not microbicidal.

a. Refrigeration temperature: stored at a temperature of 4°C to 7°C in Refrigerators.

b. Deep freezing: stored at a temperature of -20°C to -70°C in Deep Freezers.

c. Liquid Nitrogen: stored at a temperature of -196°C.

(i) Refrigeration

Refrigeration will slow down and inhibit the growth of most microbes but it will not kill them! Note: some spoilage germs and psychrophiles can continue to replicate at cooler temperatures. Organisms can be maintained viable at -80° C if suspended in glycerol.

(ii) Desiccation

Desiccation of microbes is a very useful means of food preservation and to control the growth of spoilage germs and pathogens. Foods that have a high water activity are most subject to spoilage and typically must be refrigerated or frozen. Numerous foods are preserved by adding salt or sugar to decrease the water activity of the foods. This process creates hypertonic conditions and causes water to leave bacterial cells (plasmolyze). Salting of foods does not protect against all potential pathogens. Many fungi are halophilic as is *Staphylococcus aureus*, a common source of bacterial food poisoning.

3. Radiation

One of the most controversial areas of microbial control involves the use of radiation. The controversy largely results from a lack of understanding of the different types and uses of radiation. The effects of types of radiation depend on three important factors: Time (of exposure), Distance (from the source) , Shielding (how penetrating is the radiation). Irradiation of various food has been used in the U.S since the 1960's and has been used to sterilize foods such as herbs and spices.

Microwaves, ultra violet rays, gamma rays and electrons have the power to kill the microorganisms. Among them, UV light and gamma rays are commonly used for sterilization.

(i) Nonionizing radiation

Includes microwaves and ultra violet radiation. Microwaves are not particularly antimicrobial in and of themselves, rather the killing effect of microwaves are largely due to the heat that they generate. Microwaves are not recommended for cooking large volumes or thick cuts of meat as the heat may not penetrate the foodstuffs sufficiently. Ex: UV rays.

UV rays

UV radiation is of short wavelength, between 220 and 300 nm and is not very penetrating. UV can be stopped by glass, a sheet of paper, or the top layers of your skin! UV rays can kill exposed microbes by causing damage to their DNA. UV radiation is useful for the disinfection of exposed surfaces such as laboratory hoods. However, the usefulness of UV radiation is limited by the fact that certain microbes possess DNA repair mechanisms and can recover after exposure to this kind of radiation. In addition, UV light does not penetrate organisms well that are protected in mucus or debris.

Instrument used: Laminar air flow chamber – A clean bench used to handle the microbes in a microbe free environment. It has two methods of sterilization. Radiation by UV and filtration by HEPA filters (High efficiency particulate filters), a microbe free air will be present in side.

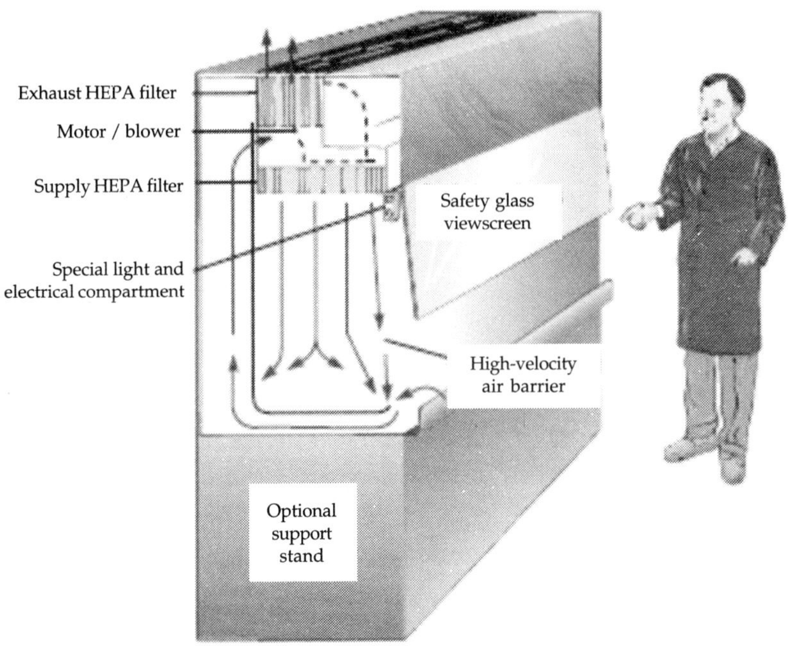

Laminar safety cabinet

(ii) Ionizing radiation

Includes gamma rays and X rays which are highly penetrating to cells and tissues and have potent antimicrobial effects. After colliding with a target, ionizing radiation generates ions and other reactive species from molecules including hydroxyl (free) OH- radicals. These free radicals can cause irreversible breaks in DNA, proteins and enzymes. Ex: Gamma rays.

Gamma Rays

Ionizing radiation is electromagnetic radiation of sufficient energy to produce ions like electron, hydroxyl ions, protons, etc. which can alter the biopolymers such as DNA and protein causing break of polymers. γ-rays produced from ^{60}Co and ^{137}Cs are useful for sterilization of the plastic wares, gloves, heat labile plastic containers, poly bags, tissue grafts, plastic syringes etc. In few cases, foods are also sterilized by g-rays.

Radiation is currently used for sterilization by the medical supply and food industries. Irradiation is the only effective means known to eliminate E. coli 0157 from meat. This process can also eliminate the food pathogens, *Listeria* , *Campylobacter* and *Salmonella.*

4. Filtration

Heat is the most common and effective way of sterilization but some heat sensitive liquids and chemicals cannot be sterilized by heat. An alternate technique for the sterilization is by filters. A filter is too small for a passage of microorganisms but large enough to passage of liquids.

Filter sterilization is commonly employed for substances that can not tolerate heat. Membrane filters with pore sizes between 0.2-0.45 um are commonly used to remove particles from solutions that can't be autoclaved. Membrane filtration of beer eliminates spoilage germs and pasteurization is no longer needed. Filtered beer is permitted to be sold as "draft beer." Submicron filters are also being marketed for removal of protozoan cysts from drinking water.

There are three types of filters

1. *Pyrex glass or sintered glass filters*: Filter glass made with powdered glass disc with pore size of 0.5 to 2 mm are used as filters. These discs are fit into funnels and used for filtration. These are used by suction pumps for faster flow rate.
2. *Seitz filters*: The filter mat is made up of asbestos – cellulose mixture.

Membrane filters

3. *Membrane filters*: They are composed of polymers with high tensile strength such as cellulose acetate, cellulose nitrate or polysulfonate, manufactured in such a way that they contain a large number of tiny holes. By adjusting the polymerization conditions during manufacture, the size of holes in the membrane can be precisely controlled. They trap the microbes on the surface and allow the microbe free solution to pass.

Application of physical agents for controlling microorganisms

Method	Recommended uses	Limitations
Moist heat Autoclave	Sterilizing instruments, linens utensils, and treatment trays, media and other liquids	Infective against organisms in materials impervious to steam : cannot be used for heat sensitive articles
Free flowing steam (or) boiling water	Destruction of nonspore forming pathogens: sanitizes bedding, clothing and dishes	Cannot be guaranteed to produce sterilization on one exposure
Dry heat Hot air oven	Sterilizing materials impermeable to or damaged by moisture, e.g., oils, glass, sharp instruments, metals	Destructive to materials which cannot withstand high temperatures for long periods
Incineration	Disposal of contaminated objects that can not be reused.	Size of incinerator must be adequate to burn largest load promptly and completely: potential of air pollution.
Radiation Ultra violet	Control of airborne infection: disinfection of surfaces	Must be absorbed to be effective (does not pass through transparent glass or opaqueobjects) irritating to eyes and skin: low penetration
X-ray ,gamma and cathode radiation	Sterilization of heat sensitive surgical materials and other medical devices	Expensive and requires special facilities for use
Filtration Membrane filters	Sterilization of heat sensitive biological fluids	Fluid must be relatively free of suspended particulate matter
Fibre glass filters(HEPA)	Air disinfection	Expensive

Chemical agents for the control of microorganisms

Although objects are sometimes disinfected with physical agents, chemicals are more often employed in disinfection and antisepsis. Many groups of chemicals are able to inhibit the growth and metabolic activity of microorganisms. An efficient chemical agent should satisfy maximum of the following characters.

Characters of chemical antimicrobial agents

1. *Antimicrobial activity* : At low concentration it should have a wider antimicrobial activity
2. *Solubility* : It must be easily soluble
3. *Stability* : Stable in all formulations
4. *Homogeneity* : Uniform in composition
5. *Toxic to microorganisms at room temperature* : No special conditions are required for the use of this agents
6. *Capacity to penetrate* : Penetration is more important for the successful germicidal action
7. *Deodorizing* : Should not produce any odour
8. *Nontoxic to humans* : Not toxic against humans

Mode of action - Antimicrobial agents

- Damage to the cell wall or inhibition of cell wall synthesis
- Alteration of the permeability of the cytoplasmic membrane
- Alteration of the physical / chemical state of proteins, nucleic acids
- Inhibition of enzyme synthesis
- Inhibition of protein or nucleic acid synthesis

Classes of chemicals involved in sterilization

Different chemicals like halogens, heavy metals, phenols, alcohols and ethanol are used for sterilization.

1. Phenols & phenolic compounds

Phenol was the first widely used antiseptic and disinfectant. In 1867 Joseph Lister employed it to reduce the risk of infection during operations. Today phenol and phenolics (phenol derivatives) such as cresols, xylenols, and orthophenylphenol are used as disinfectants in laboratories and hospitals. The commercial disinfectant Lysol is made of a mixture of phenolics.

Phenolics act by denaturing proteins and disrupting cell membranes. They have some real advantages as disinfectants: phenolics are tuberculocidal, effective in the presence of organic material, and remain active on surfaces long after application. However, they do have a disagreeable odor and can cause skin irritation.

2. Alcohols

Alcohols are among the most widely used disinfectants and antiseptics. They are bactericidal and fungicidal but not sporicidal; some lipid-containing viruses are also destroyed. The two most popular alcohol germicides are ethanol and isopropanol, usually used in about 70 to 80% concentration. They act by denaturing proteins and possibly by dissolving membrane lipids. 10 to 15 minute soaking is sufficient to disinfect thermometers and small instruments.

3. Halogens

A halogen is any of the five elements (fluorine, chlorine, bromine, iodine, and astatine) in group VIIA of the periodic table. They exist as diatomic molecules in the free state and form saltlike compounds with sodium and most other metals. The halogens iodine and chlorine are important antimicrobial agents.

Iodine is used as a skin antiseptic and kills by oxidizing cell constituents and iodinating cell proteins. At higher concentrations, it may even kill some spores. Iodine often has been applied as tincture of iodine, 2% or more iodine in a water-ethanol solution of potassium iodide. Although it is an effective antiseptic, the skin may be damaged, a stain is left, and iodine allergies can result. More recently iodine has been complexed with an organic carrier to form an *iodophor*. Iodophors are water soluble, stable, and nonstaining, and release iodine slowly to minimize skin burns and irritation. They are used in hospitals for preoperative skin degerming and in hospitals and laboratories for disinfecting. Some popular brands are Wescodyne for skin and laboratory disinfection and Betadine for wounds.

Chlorine is the usual disinfectant for municipal water supplies and swimming pools and is also employed in the dairy and food industries.

It may be applied as chlorine gas, sodium hypochlorite, or calcium hypochlorite, all of which yield hypochlorous acid (HClO) and then atomic oxygen. The result is oxidation of cellular materials and destruction of vegetative bacteria and fungi, although not spores. Chlorine is also an excellent disinfectant for individual use because it is effective, inexpensive, and easy to employ. Small quantities of drinking water can be disinfected with halazone tablets. Halazone (parasulfone dichloramidobenzoic acid) slowly releases chloride when added to water and disinfects it in about a half hour. It is frequently used by campers lacking access to uncontaminated drinking water. Chlorine solutions make very effective laboratory and household disinfectants.

4. Heavy metals

For many years the ions of heavy metals such as mercury, silver, arsenic, zinc, and copper were used as germicides. More recently these have been superseded by other less toxic and more effective germicides (many heavy metals are more bacteriostatic than bactericidal).

There are a few exceptions. 1% solution of silver nitrate is often added to the eyes of infants to prevent ophthalmic gonorrhea (in many hospitals, erythromycin is used instead of silver nitrate because it is effective against *Chlamydia* as well as *Neisseria*). Silver sulfadiazine is used on burns. Copper sulfate is an effective algicide in lakes and swimming pools.

Heavy metals combine with proteins, often with their sulfhydryl groups, and inactivate them. They may also precipitate cell proteins.

5. Dyes

Commonly used dyes are Triphenyl Methane and Acridine orange. Gram positive are more sensitive to dyes. Dyes are used in selective media preparation. Eg. Brilliant green, Crystal violet inhibit Gram positive

6. Quaternary Ammonium compounds (Quaternary disinfectants)

Detergents [Latin *detergere,* to wipe off or away] are organic molecules that serve as wetting agents and emulsifiers because they have both polar hydrophilic and nonpolar hydrophobic ends. Due to their amphipathic nature, detergents solubilize otherwise insoluble residues and are very effective cleansing agents. They are different than soaps, which are derived from fats.

Although anionic detergents have some antimicrobial properties, only cationic detergents are effective disinfectants. The most popular of these disinfectants are quaternary ammonium compounds characterized by a positively charged quaternary nitrogen and a long hydrophobic aliphatic chain.They disrupt microbial membranes and may also denature proteins.

Cationic detergents like benzalkonium chloride and cetylpyridinium chloride kill most bacteria but not *M. tuberculosis* or endospores. They do have the advantages of being stable, nontoxic, and bland but they are inactivated by hard water and soap. Cationic detergents are often used as disinfectants for food utensils and small instruments and as skin antiseptics. Several brands are on the market. Zephiran contains benzalkonium chloride and Ceepryn, cetylpyridinium chloride.

7. Aldehydes

Both of the commonly used aldehydes, formaldehyde and glutaraldehyde, are highly reactive molecules that combine with nucleic acids and proteins and inactivate them, probably by crosslinking and alkylating molecules. They are sporicidal and can be used as chemical sterilants. Formaldehyde is usually dissolved in water or alcohol before use. 2% buffered solution of glutaraldehyde is an effective disinfectant. It is less irritating than formaldehyde and is used to disinfect hospital and laboratory equipment. Glutaraldehyde usually disinfects objects within about 10 minutes but may require as long as 12 hours to destroy all spores.

8. Gaseous agents

Many heat-sensitive items such as disposable plastic Petri dishes and syringes, heart-lung machine components, sutures, and catheters are now sterilized with ethylene oxide gas.

Ethylene oxide (EtO) is both microbicidal and sporicidal and kills by combining with cell proteins. It is a particularly effective sterilizing agent because it rapidly penetrates packing materials, even plastic wraps. Sterilization is carried out in a special ethylene oxide sterilizer, very much resembling an autoclave in appearance, that controls the EtO concentration, temperature, and humidity. Because pure EtO is explosive, it is usually supplied in a 10 to 20% concentration mixed with either CO_2 or dichlorodifluoromethane. The ethylene oxide concentration, humidity, and temperature influence the rate of sterilization. A clean object can be sterilized if treated for 5 to 8 hours at 38°C or 3 to 4 hours at 54°C when the relative humidity is maintained at 40 to 50% and the EtO concentration at 700 mg/liter. Extensive aeration of the sterilized materials is necessary to remove residual EtO because it is so toxic.

Betapropiolactone (BPL) is occasionally employed as a sterilizing gas. In the liquid form it has been used to sterilize vaccines and sera. BPL decomposes to an inactive form after several hours and is therefore not as difficult to eliminate as EtO. It also destroys microorganisms more readily than ethylene oxide but does not penetrate materials well and may be carcinogenic. For these reasons, BPL has not been used as extensively as EtO. Recently vapor-phase hydrogen peroxide has been used to decontaminate biological safety cabinets.

Testing the efficiency of chemical agents

Testing of antimicrobial agents is a complex process regulated by two different federal agencies. The U.S. Environmental Protection Agency regulates disinfectants, whereas agents used on humans and animals are under the control of the Food and Drug Administration. Testing of antimicrobial agents often begins with an initial screening test to see if they are effective and at what concentrations. This may be followed by more realistic in-use testing.

The best-known disinfectant screening test is the *phenol coefficient test* in which the potency of a disinfectant is compared with that of phenol. The ratio between the greatest dilution of a test germicide capable of killing a test organism in 10 min. but not in 5 min. and the greatest dilution of phenol giving the same result is called phenol coefficient. The phenol coefficient test is a useful initial screening procedure, but the phenol coefficient can be misleading if taken as a direct indication of disinfectant potency during normal use. This because the phenol coefficient is determined under carefully controlled conditions with pure bacterial strains, whereas disinfectants are normally used on complex populations in the presence of organic matter and with significant variations in environmental factors like pH, temperature, and presence of salts.

To more realistically estimate disinfectant effectiveness, other tests are often used. The rates at which selected bacteria are destroyed with various chemical agents may be experimentally determined and compared. A *use dilution test* can also be carried out. Stainless steel cylinders are contaminated with specific bacterial species under carefully controlled conditions. The cylinders are dried briefly, immersed in the test disinfectants for 10 minutes, transferred to culture media, and incubated for two days. The disinfectant concentration that kills the organisms in the sample with a 95% level of confidence under these conditions is determined.

Disinfectants also can be tested under conditions designed to simulate normal in-use situations. In-use testing techniques allow a more accurate determination of the proper disinfectant concentration for a particular situation.

Common antibiotics and their mode of action

S. No.	Mode of action	Antibiotics	Source	Effective against
1.	Inhibit the cell wall (Peptidoglycan layer) synthesis	Penicillin	*Penicillium notatum, P. crysogenum*	Gram +ve bacteria
		Ampiicillin	penicillin	Gram +ve bacteria
		Cephalosporins	*Cephalosporium acremonium*	G +ve, G-ve
		Cycloserine	*Streptomyces* sp Now chemical synthesis	Gram +ve bacteria
		Bacitracin	*Bacillus subtilis*	G +ve
		Vancomycin	*Streptomyces orientalis*	G +ve
2.	Damage the cytoplasmic membrane	Polymyxin	*Bacillus* spp	G-ve
		Gramicidin	*Bacillus* spp	G+ve
		Nystatin	*Streptomyces noursei*	Fungus, yeast
3.	Inhibition of protein synthesis	Streptomycin	*Streptomyces griseus*	G + and G- bacteria
		Chlortetracycline	*Streptomyces aureofaciens*	G + and G- bacteria
		Oxytetracyclin	*Streptomyces rimosus*	
		Chloramphenicol	*Streptomyces venezuelae*	G + and G- bacteria
		Erythromycin	*Streptomyces erythraeus*	G + and G- bacteria

Common chemicals used as disinfectants and antiseptics

Chemical	Use	Mode of action
Antisepsis		
Alcohol	Skin	Lipid solvent and protein denaturant
Phenol compounds	Soaps, lotions, cosmetics, body deodorants	Disrupt cell membrane
Cationic detergents (Ex. Benzalkonium chloride)	Soap, lotion	Pospholipid interaction
3% hydrogen peroxide	Skin	Oxidizing agent
Iodine containing iodophors	Skin	Iodinates tyrosine residues; oxidizing agent
Silver nitrate	Eyes of new born to prevent blindness from *Nesseria gonorrhoea*	Protein precipitant
Disinfectants		
Alcohol	Medical instruments, food, dairy instruments, lab surface	Lipid solvent & protein denaturant
Cationic detergents	Medical instruments, food, dairy instruments, lab surface	Interact with phospholipids
Chlorine gas	Disinfectant for purification of water supplies	Oxidizing agent
Chlorine compounds (Chloramines, sodium hypochlorite, chlorine dioxide)	Disinfectant for food and dairy industry, water supplies	Oxidizing agent
Copper sulphate	Algicide in swimming pools	Protein precipitant
Ethylene oxide (gas)	Sterilant for temperature sensitive lab material like plastics	Alkylating agent
Formoldehyde	3-8% used as surface disinfectant; 37% (referred as formalin) as preservatives	Alkylating agent

Gluteraldehyde	2% solution	Alkylating agent
3% hydrogen peroxide	Medical instruments	Oxidizing agent
Mercuric chloride	For laboratory surface, plant samples	Protein denaturation
Ozone	Drinking water	Oxidizing agent
Preservatives		
Salt and sugar solutions	Preservative	Osmotic pressure

□□□

Chapter 11

Microbial Metabolism

Basic Concepts of Metabolism

Before going into the microbial metabolism, it is essential to understand the principles of bioenergetics.

Principles of Bioenergetics

Energy

It is defined as the capacity to do work. The energy can be referred as kcal or K Joules. One kcal is the energy required to raise 1°C of 1 kg of water. 1 calorie = 4.184 Joules

Most of the cells obtain energy by carrying out the chemical reactions viz., oxidation of organic or inorganic compounds which liberate energy. The energy released during oxidation or reduction reaction is called chemical energy.

The living things should take the energy from existing sources and convert them to a suitable form for the biological process.

Plants absorb the energy from light and convert them to chemical compounds of high energy. Similarly, animals can derive their energy by oxidizing the chemicals. Microorganisms especially the bacteria have both the ways (from chemicals and sun light)

In any chemical reaction, energy will be liberated or absorbed. The released energy is called total energy.

Free Energy (ΔG)

Free energy (ΔG) is defined as the part of the total energy which is available for the performance of some useful work.

Entrophy

Entrophy is a measure of the part of the total energy which is unavailable for the useful work.

Total energy = free energy (ΔG) + entropy

- If ΔG of a chemical reaction has a negative value (- 5000 cal), the reaction is exergonic reaction.
- If ΔG of a chemical reaction has a positive value (2500 cal), the reaction requires energy (*i.e.* Endergonic reaction).

A reaction may release or require free energy. If a reaction needs energy, it is referred as *endergonic reaction* and if a reaction releases energy, it is referred as *exergonic reaction.*

A + B → C + D + energy [exergonic reaction]

A + B + energy → C + D [endergonic reaction]

ΔG of some compounds are,

H_20 = -237.2; CO_2 = -396.4; H_2 = 0; O_2 = 0; N_2O = +104.2; Glucose = -917.3.

Negative ΔG is good and available to do work in the cell.

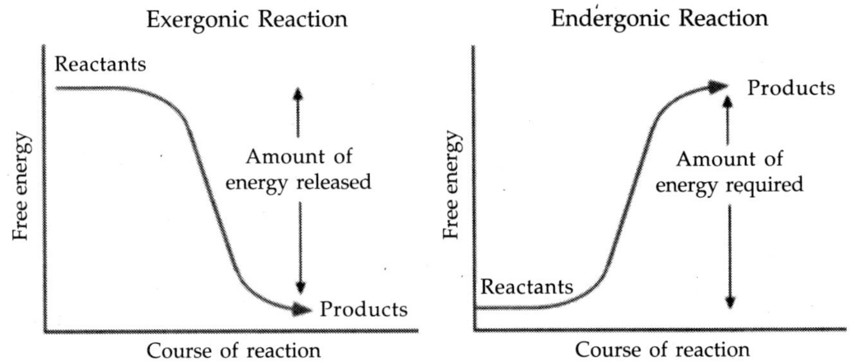

Exergonic and endergonic reactions

Activation energy

It is referred as the energy required to bring all the molecules to reactive stage.

When the H_2 and O_2 gases kept together in a container will not react each other even after many years and need some energy to make them as reactive molecules to react and gives H_2O. That energy is referred as activation energy.

Catalyst

Catalyst is a substance to lower the activation energy requirement of a reaction and also to increase the reaction rate.

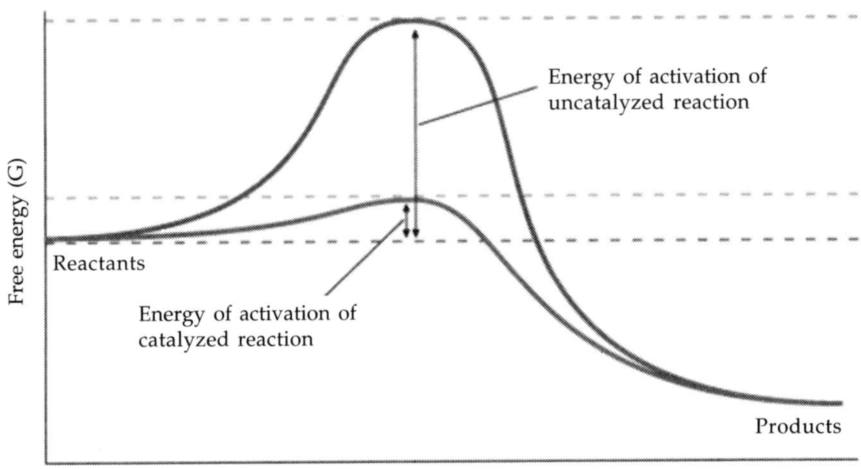

Above diagram shows that catalyst lowers the activation energy

In biological systems, proteins involved in the catalytic activity are referred as **biocatalysts** or **enzymes**. They increase the reaction rate of 10^8 to 10^{20} times.

Enzymes

Enzymes are protein in nature with the size of 1×10^4 to 1×10^6 Dalton (1 Dalton $=1.67 \times 10^{-24}$ g). Normally some non-proteinaceious, low molecular substances were joint together with protein part to serve as enzymes. Enzymes are very specific, efficient and can be inhibited by products.

Protein + Non-protein compounds → enzyme

The protein part is *Apoenzyme;* The non-protein is referred as *Co-enzyme* or *Prosthetic group.*

The both (apoenzyme and coenzyme/prosthetic group) are referred as *Holoenzymes.*

The difference between co-enzyme and prosthetic group is that co-enzyme will have loose bound with apoenzyme (temporarily) whereas the prosthetic group will have tight bound with apoenzymes (permanently).

For Example,

- Co-enzymes – NADP, NAD, FAD, FAM
- Prosthetic group – Heme group of cytochromes

Apart from the co-enzymes and prosthetic groups, the inorganic ions like Ca, Mg, Fe, Mn are required for the activation of enzyme and they are referred as *co-factors.*

Role of ATP in Metabolism

Energy is released from a cell in an exergonic reaction (i.e. The reactions with – ΔG). Instead of wasting this energy, much of it is trapped and used for doing work (endergonic reactions). In living organisms this practical form of energy is adenosine triphosphate (ATP). In a sense, cells carry out certain processes so that they can earn ATP and carry out other processes in which they spend their ATP. Thus, ATP is referred to as the cell's energy currency. ATP serves as the link between exergonic reactions and endergonic reactions. ATP is a high energy molecule. It breaks down or hydrolyses completely to the products adenosine diphosphate (ADP) and orthophosphate + energy. This energy is used to resynthesize ATP from ADP and P_i during catabolism. Thus, ATP, ADP and P_i form an energy cycle.

Energy Yielding Reactions

1. Oxidation – Reduction reaction

Several types of chemical reactions are involved in energy production. Oxidation – reduction reaction is the common method to derive energy by the living cells.

Oxidation is the loss of electrons and *reduction* is the gain of electrons.

Oxidation reduction reactions are dehydrogenations (reaction involves the loss of hydrogen atoms).

(Note: Remember the phrase "LEO says GER" LEO – Loss of Electron is Oxidation; GER – Gain of Electron is Reduction)

Oxidizing agent

An oxidizing agent (oxidant) will absorb electrons and will therefore become reduced, as illustrated by the following examples.

The ferric ion is an oxidizing agent; it absorbs electrons and becomes reduced to ferrous ion:

Fe^{3+}	+	e^-	→	Fe^{2+}
Ferric ion		Electron		Ferrous ion

The hydrogen ion is an oxidizing agent; it absorbs electrons and becomes reduced to atomic hydrogen:

$2H^+$	+	e^-	→	$2H$
Hydrogen ion: proton		Electron		Hydrogen atom: proton plus electron

Reducing agent

A reducing agent (*reductant*) donates electrons, becoming oxidized in the process. The ferrous ion is a reducing agent, it donates electrons and becomes oxidized to ferric ion:

Fe^{2+}	→	Fe^{3+}	+	e^-
Ferrous ion		Ferric ion		Electron

Here, the Fe^{3+} accepts electrons and hence referred as *electron acceptor*. Fe^{2+} donates the electron and is referred as *electron donor*.

Oxidation - reduction potential

Substrates or compounds vary their ability to donate or accept the electrons (in other words, to become oxidized or reduced). The tendency of a substrate to donate or accept the electron is referred as oxidation – reduction Potential or *redox potential*. The redox potential of each substance was calculated by means of electrical volt with reference substance as H_2. The redox potential is –ve means, the substance is rich of electrons, (means reduced form) and +ve means it already lost the electrons (means oxidized form). The reduced forms can give or donate electrons and oxidized form can accept the electrons.

Electron carriers

In a series of oxidation-reduction reactions, the electron from one reaction can be carried out to another through some carriers called *electron carriers*. Ex: NAD, NADP, FAD, FAM, cytochromes, etc.

2. Phosphorylation

Addition of a PO_4 group to a compound is called phosphorylation. The energy released during oxidation reaction is stored in the form of *high energy phosphate bonds* in ATP. When energy is released, the ADP and Pi reacted and form ATP. This process of addition of Pi to form ATP is referred as phosphorylation.

Three types of phosphorylation occurs through which the ATP formation occurs in bacteria.

a. *Substrate level phosphorylation* - During the oxidation of one organic molecule, energy will be released, which can be used for the production of ATP. The organic substrate itself acts as electron acceptor. This kind of phosphorylation is referred as *substrate level phosphorylation*.

b. *Oxidative phosphorylation* - During electron transport chain, a series of oxidation – reduction reactions occur in cells. These reactions are mediated by a number of enzymes. As the electron flows through the chains, much of their free energy is conserved in the form of ATP. This process is called oxidative phosphorylation. It is also referred as *electron transport phosphorylation*.

c. *Photophosphorylation* – Utilization of light energy to derive the synthesis of ATP is referred as phosphorylation. The proton motive force during photosynthesis is used for ATP synthesis.

Microbial Metabolism

Metabolism refers the sum of biochemical reactions required for energy generation and the use of energy to synthesize cellular materials.

Metabolism includes two types of reactions

- Catabolism
- Anabolism

Catabolism refers to the breakdown of a substance for generation of energy

Anabolism refers to the synthesis or build up of macromolecules and cell organelles

During catabolism, the energy is conserved as ATP and it is utilized in anabolism, which has the value of 8 kcal per mole.

I. Energy generation by heterotrophic microorganisms

Heterotrophs are the microorganisms that are unable to use carbondioxide as its sole source carbon and requires one or more organic compounds. Heterotrophic bacteria can use a variety of organic compounds as energy sources. The compounds include carbohydrates, fatty acids and amino acids. The most preferred compounds are carbohydrates especially 6 carbon sugar – Glucose.

Glucose can be taken up by bacterium through diffusion and can be readily utilized. The bacteria can use glucose by following these pathways.

a. Embden Meyerhof pathway
b. Phosphoketolase pathway
c. Entner - Doudoroff pathway
d. Pentose phosphate pathway

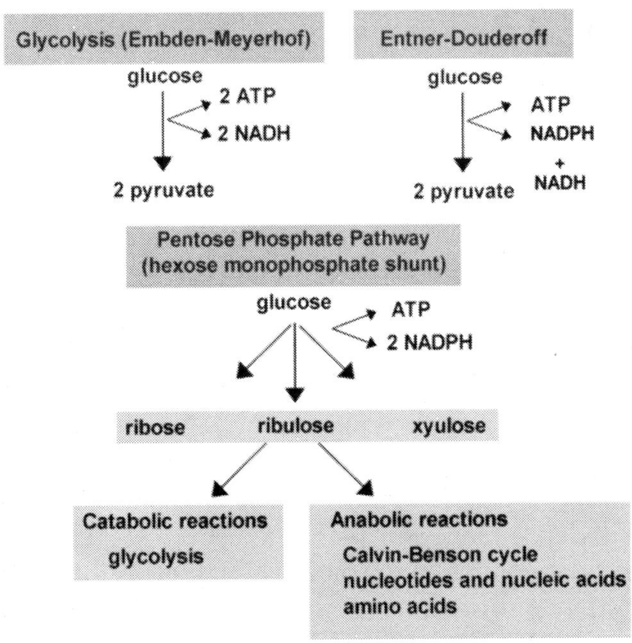

All these pathways are fermentative pathways and during these processes, substrate level phosphorylation occurs through which ATP synthesis takes place.

1. Substrate level phosphorylation

a) Embden – Meyerhof pathway

The most common pathway of glucose catabolism is the Embden-Meyerhof pathway of *glycolysis* ("splitting of glucose"). This process occurs very widely and is found in microorganisms as well as in animals and plants. Glycolysis does not require the presence of oxygen and therefore can occur in both aerobic and anaerobic cells. Aerobic cells degrade glucose by glycolysis, and this process constitutes the preparatory stage for aerobic phase of glucose oxidation.

Thus, under *anaerobic conditions* this situation prevails:

$$\text{Glucose} \xrightarrow{\text{fermentation}} \text{fermentation products}$$

Whereas under **aerobic conditions**, the following occurs:

$$\text{Glucose} \xrightarrow{\text{fermentation}} \underset{\text{(pyruvate)}}{\text{intermediate}} \xrightarrow[O_2]{\text{respiration}} CO_2 + H_2O$$

This pathway operates in yeast to produce alcohol and in lactic acid bacteria to produce lactic acid and other organic acids, fatty acids and alcohols.

The overall reaction of EMP of glycolysis is as follows:

$$C_6H_{12}O_6 + 2NAD + 2ATP + \underset{\text{Inorganic Phosphate}}{2Pi} \rightarrow \underset{\text{Pyruvic acid}}{2CH_3COCOOH} + 2NADH_2 + 2ATP$$

After pyruvate formation, if the organism is aerobic/respiratory type, it enters into the Kreb's cycle and if the organism is fermentative/ anaerobic, the reduction process occurs which ends with the production of organic acids/alcohols, etc. Complete oxidation leads to respiration and the reduction process leads to fermentation.

The Embden – meyerhof path way can lead to a wide array of end products depending on the path ways taken in the reductive steps after the pyruvate formation.

The following are some of such fermentations:

Fermentation	End products	Model organism
Homolactic fermentation	Lactic acid	*Lactobacillus*
Mixed acid fermentation	Lactate, acetate, formate, succinate	*Enterobater*
Butyric acid fermentation	Butric acid, acetone	*Clostridium acetobutylicum*
Propionic acid fermentation	Propionic acid	*Propionibacterium*
Alcohol fermentation	Ethanol	*Saccharomyces*

b) Phosphoketolase pathway / Heterolactic pathway

The phosphoketolase path way is distinguished by the key cleavage enzyme *phosphoketolase*, which cleaves pentose to glyceroldehyde 3 phosphate and acetyl phosphate. The path way ends with ethanol and lactic acid. Ex. *Lactobacillus, Leuconostoc.*

The overall reaction is

Glucose $\rightarrow$ 1 lactate + 1 ethanol + 1 CO_2 + 1 ATP

This path way is useful in the dairy industry for preparation of kefir (fermented milk), yogurt, etc.

c) Entner - Doudoroff pathway

This pathway is found in both aerobic and anaerobic prokaryotes but not in eukaryotes. It is widespread among soil microbes such as *Pseudomonas, Azotobacter, Agrobacterium* and a few Gram negative bacterium, very few Gram positive have this pathway (*Enterococcus faecalis*).

Here, glucose is phosphorylated to glucose-6-phosphate. It is then oxidized to 6-phosphogluconic acid. A dehydration step follows to yield 2-keto-3-deoxy-6-phosphogluconic acid (KDPG); the latter is cleaved to pyruvic acid and glyceraldehydes-3-phosphate, which is metabolized via some EMP pathway enzymes to produce a second molecule of pyruvic acid. In the aerobic pseudomonads, the catabolism is completed via acetyl - CoA and the tricarboxylic acid cycle.

This pathway will be operating in *Zymomonas mobilis* by converting pyruvate to ethanol.

Glucose $\rightarrow$ 2 ethanol + 2 CO_2 + 1 ATP

The Enter-Doudoroff pathway yields one ATP, one NADPH, and one NADH molecule from one glucose molecule.

Used by Gram negatives (e.g. *Rhizobium, Psuedomonas, Agrobacterium*), usually not by Gram positives.

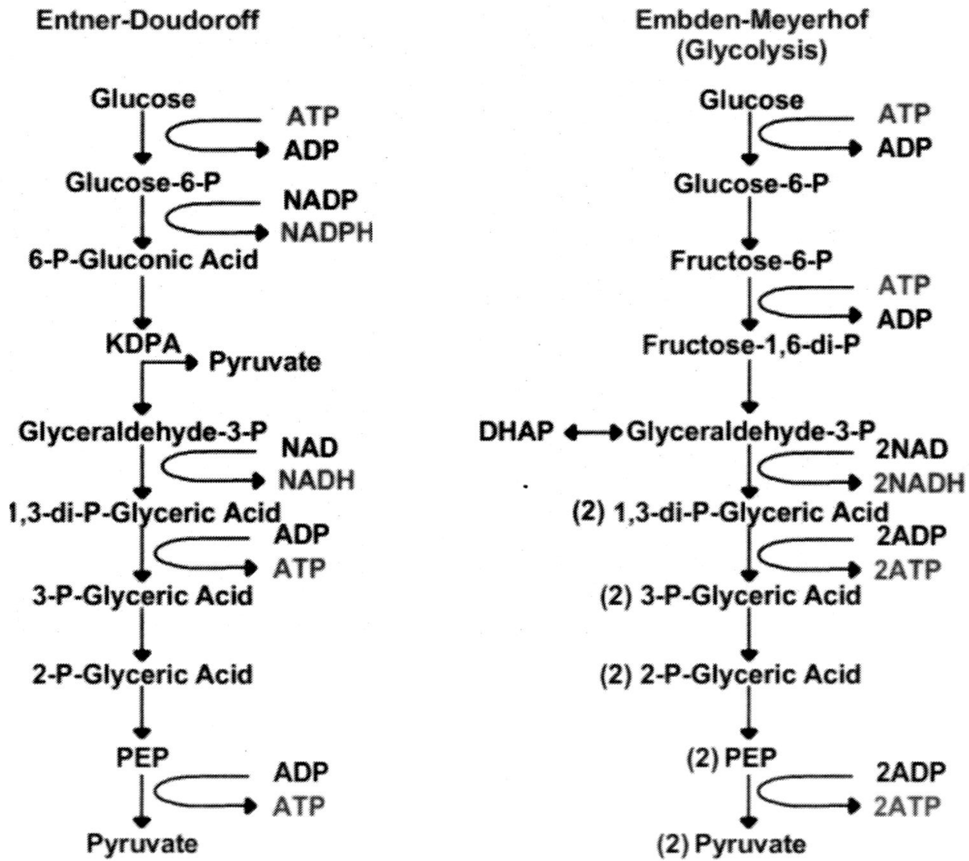

d) Pentose Phosphate Pathway

The pentose phosphate pathway like the glycolytic one, is another catabolic reaction pathway that exists in both prokaryotic and eukaryotic cells. Since it involves some reactions of the glycolytic pathway, it has been viewed as a 'shunt' of glycolysis. Hence it may be called as pentose phosphate pathway or phosphogluconate pathway. Glucose can be oxidized by the pentose phosphate pathway with the liberation of electron pairs, which may enter the respiratory chain. However this cycle is not generally considered a major energy yielding pathway in most organisms. It provides reducing power in the form of NADPH + H^+, which is required in many biosynthetic reactions of the cell, and it provides pentose phosphates for use in nucleotide synthesis. Although it can produce energy for the cell as an alternate pathway for the oxidation of glucose, it is also a mechanism for obtaining energy from 5-carbon sugars.

II. Oxidative phosphorylation (respiration)

It is one of the energy generative pathway in aerobic organisms. If the organism is a respiratory type, it needs some of the essential components for their respiration and oxidative phosphorylation to occur.

a. *Tricarboxylic acid cycle (also known as citric acid cycle or Kreb's cycle)* The pyruvate formed during glycolysis will be completely oxidized to 3 CO_2 by the use of this cycle. During oxidation of one pyruvate through TCA cycle, 4 $NADH_2$,1 $FADH_2$ and 1 GTP are produced along with 3 CO_2.

b. *A membrane and associated Electron Transport System (ETC)* The electron transport chain is a sequential transfer of electrons through electron carriers. During this flow of electron in the membrane, a proton motive force across the membrane leads to the formation of ATP (is referred as electron transport phosphorylation).

c. *An outside electron carrier*: for aerobic respiration, O_2 serves as the terminal electron acceptor and get reduced to H_2O. This is normal in higher organisms. But in anaerobic bacteria, the terminal electron acceptor may be of nitrite, nitrate, sulphate or carbon dioxide.

d. *A membrane bound ATPase enzyme*: The proton motive force developed during ETC leads to the formation of ATP by enzyme ATPase present in the membrane.

The Pentose Phosphate pathway (hexose monophosphate shunt) is used to metabolize five-carbon sugars; **one ATP** and **2 NADPH** molecules are produced from oxidation of one glucose molecule.

Produces intermediates for nucleotide and nucleic acid synthesis, glucose synthesis from CO_2 in photosynthesis, and some amino acids.

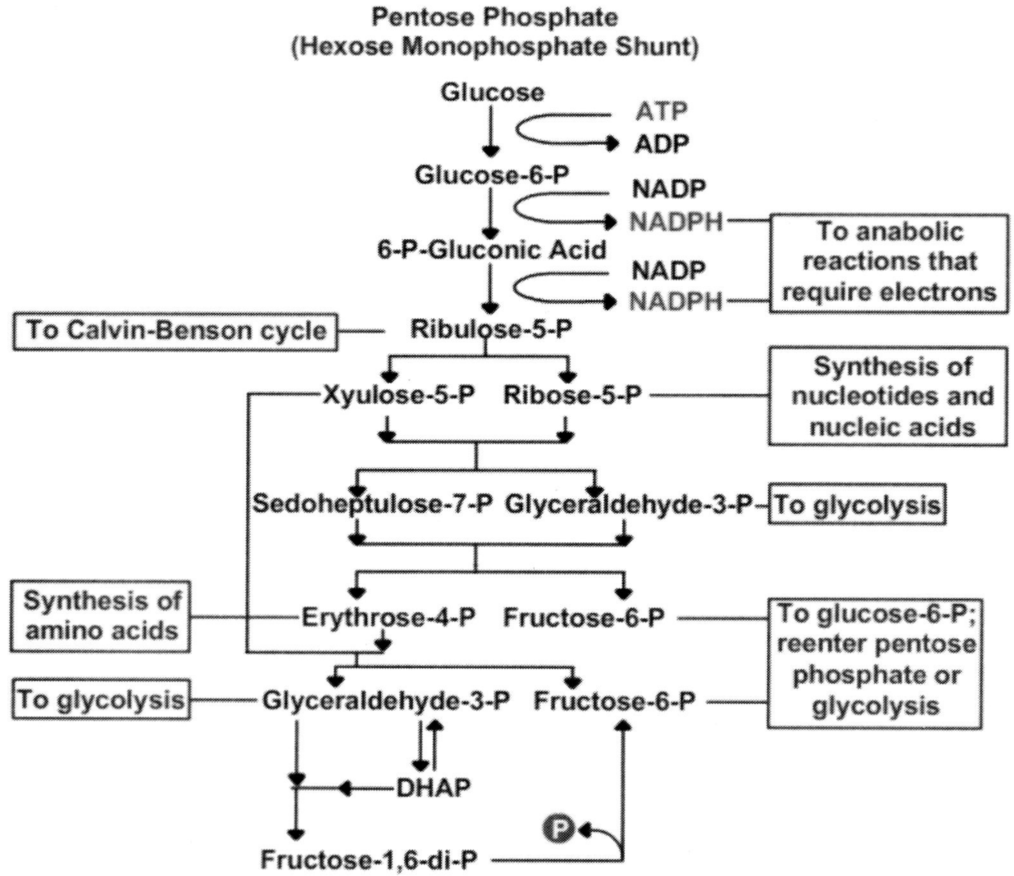

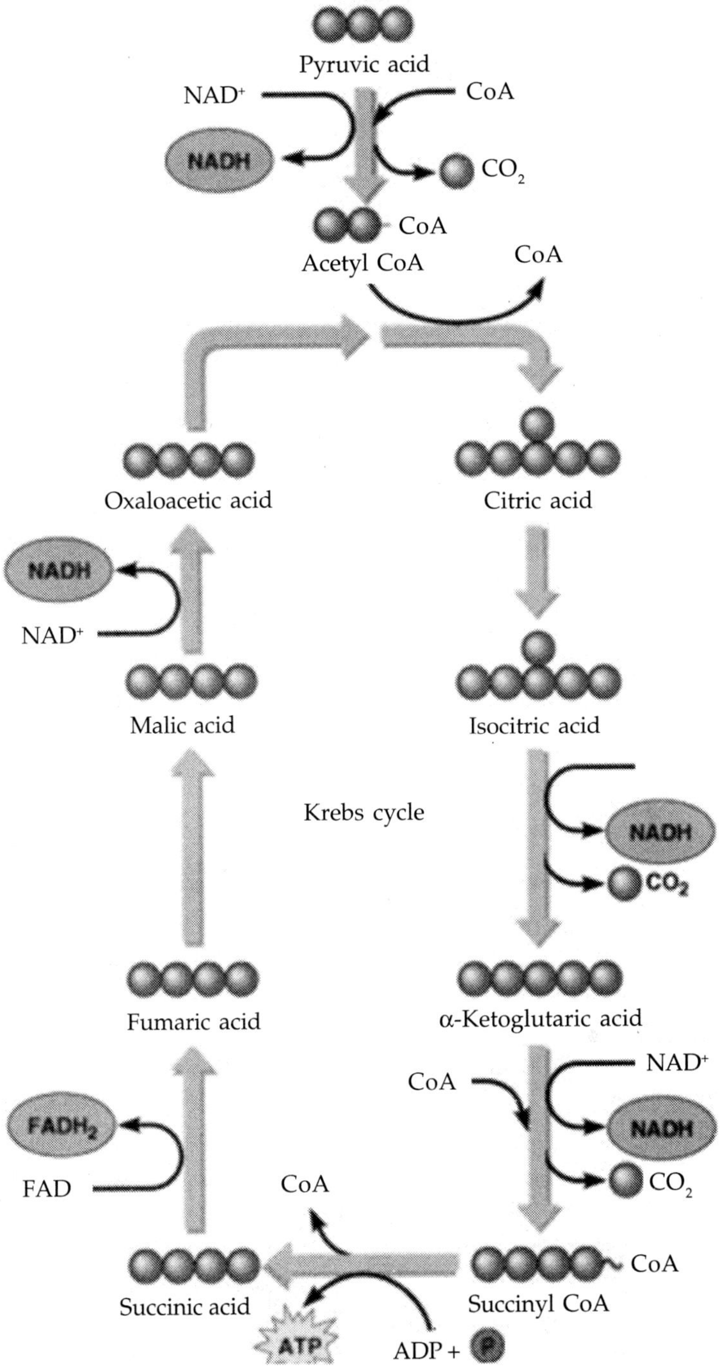

The TCA cycle

TCA cycle / Kreb's cycle / citric acid cycle

The tricarboxylic acid (TCA) cycle is a sequence of reactions that generates energy in the form of ATP and reduced coenzyme molecules ($NADH_2$ and $FADH_2$). It also performs other functions. Many intermediates in the cycle are precursors in the biosynthesis of amino acids, purines, pyrimidines, etc. For example, oxaloacetic acid and α – ketoglutaric acid are amino acid precursors.

Thus the TCA cycle is an *amphibolic cycle,* which means that it functions not only in catabolic (breakdown) but also in anabolic (synthesis) reactions.

The overall reaction of the TCA cycle can be summarized as follows:

Acetyl - CoA + 3 H_2O + 3 NAD^+ + FAD + ADP + $P_i \rightarrow 2CO_2$ +CoA + $3NADH_2$ +$FADH_2$ +ATP

Since the breakdown of glucose by glycolysis yields two acetyl - CoA molecules which can enter this cycle, the overall equation for the cycle, per glucose molecule breakdown, is twice as above.

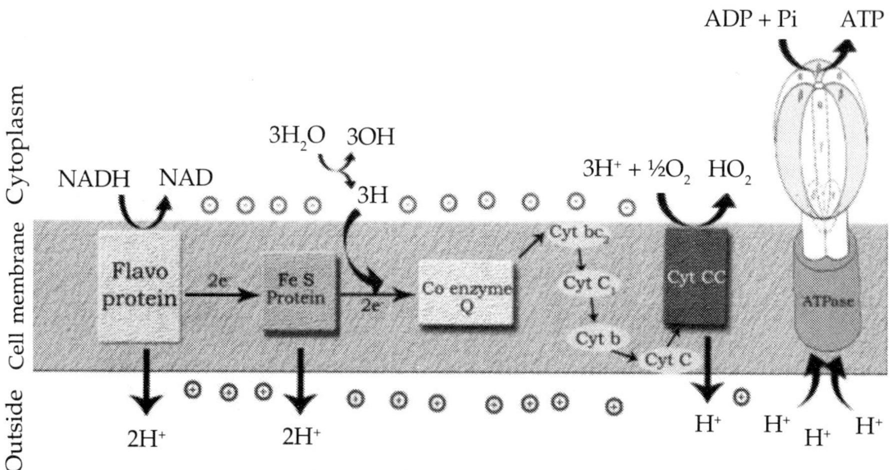

Electron Transport Chain

Energy yield in aerobic respiration

We may now look at the energy yield from the aerobic breakdown of some molecules of glucose when the electrons stored in the reduced coenzyme molecules are fed into the electron - transport chain. As shown previously, the electrons are transferred stepwise from the coenzyme carriers to molecular oxygen, and this transfer is coupled to the generation of ATP by oxidative phosphorylation

For each glucose molecule broken down, there are 12 reduced coenzymes to be oxidized: 2 $FADH_2$ (1 from each turn of the TCA cycle) and 10$NADH_2$ (2 from glycolysis; 2 from the gateway step between glycolysis and the TCA cycle, i.e., pyruvic acid to acetyl - CoA; and 6 from two turns of the TCA cycle). Since 3 ATP are produced from each $NADH_2$ and 2 ATP from each $FADH_2$, there are 34 ATP generated from the reduced coenzymes via oxidative phosphorylation through the respiratory chain. But the total yield of ATP from the aerobic respiration of 1 glucose molecule is 38: 34 from the oxidation of reduced coenzymes, 2 from glycolysis, and 2 from the side reaction of TCA cycle, that is, from 2 GTP.

The complete oxidation of glucose via glycolysis, the TCA cycle, and the respiratory chain is summarized in this overall reaction:

$$C_6H_{12}O_6 + 6O_2 \rightarrow 6CO_2 + 6H_2O$$

Anaerobic respiration

Some bacteria which utilizes reduced inorganic compounds as final electron acceptor as the substitute for oxygen is called anaerobic respiration.

The pathways for the dissimilation of the carbon and energy sources are identical with those in aerobic respiration, and electron transport occurs via a respiratory chain similar to that in aerobic cells. Oxygen is replaced as the terminal electron acceptor by nitrate. However, in some strict anaerobes, other compounds, such as carbon di oxide, or ions, such as sulphate ion, can be the terminal electron acceptors.

II. Energy generation by autotrophs

Autotrophs are self synthesizing and self nourishing organisms. Based on the source of energy and carbon there are two types of autotrophs.

a) Photoautotrophs – Use light as energy source and CO_2 as carbon source.

b) Chemoautotrophs – Use chemicals, specially the reduced inorganic compounds as C source.

The following contents will describe how the energy is produced and assimilated in autotrophs.

Photoautotrophy (Carbon assimilation by phototrophs)

Source		ATP Yield (Method)
Glycolysis		
1. Oxidation of glucose to pyruvic acid	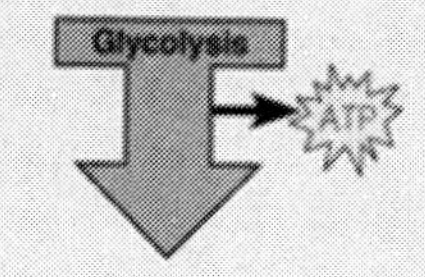	2 ATP (substrate-level phosphorylation)
2. Production of 2 NADH		6 ATP (oxidative phosphorylation in electron transport chain)
Preparatory Step		
1. Formation of acetyl CoA produces 2 NADH	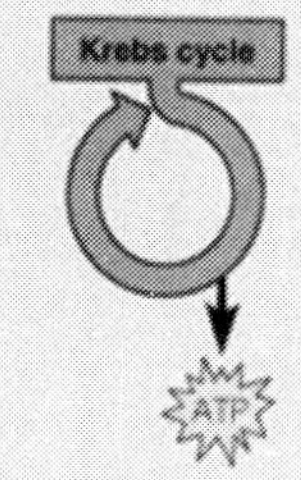	6 ATP (oxidative phosphorylation in electron transport chain)
Krebs Cycle		
1. Oxidation of succinyl CoA to succinic acid		2 GTP (equivalent of ATP; substrate-level phosphorylation)
2. Production of 6 NADH		18 ATP (oxidative phosphorylation in electron transport chain)
3. Production of 2 FADH		4 ATP (oxidative phosphorylation in electron transport chain)
		Total: 38 ATP

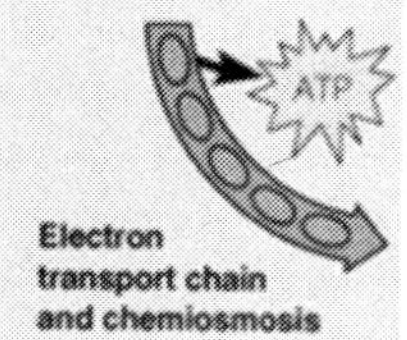

ATP generation in aerobic respiration

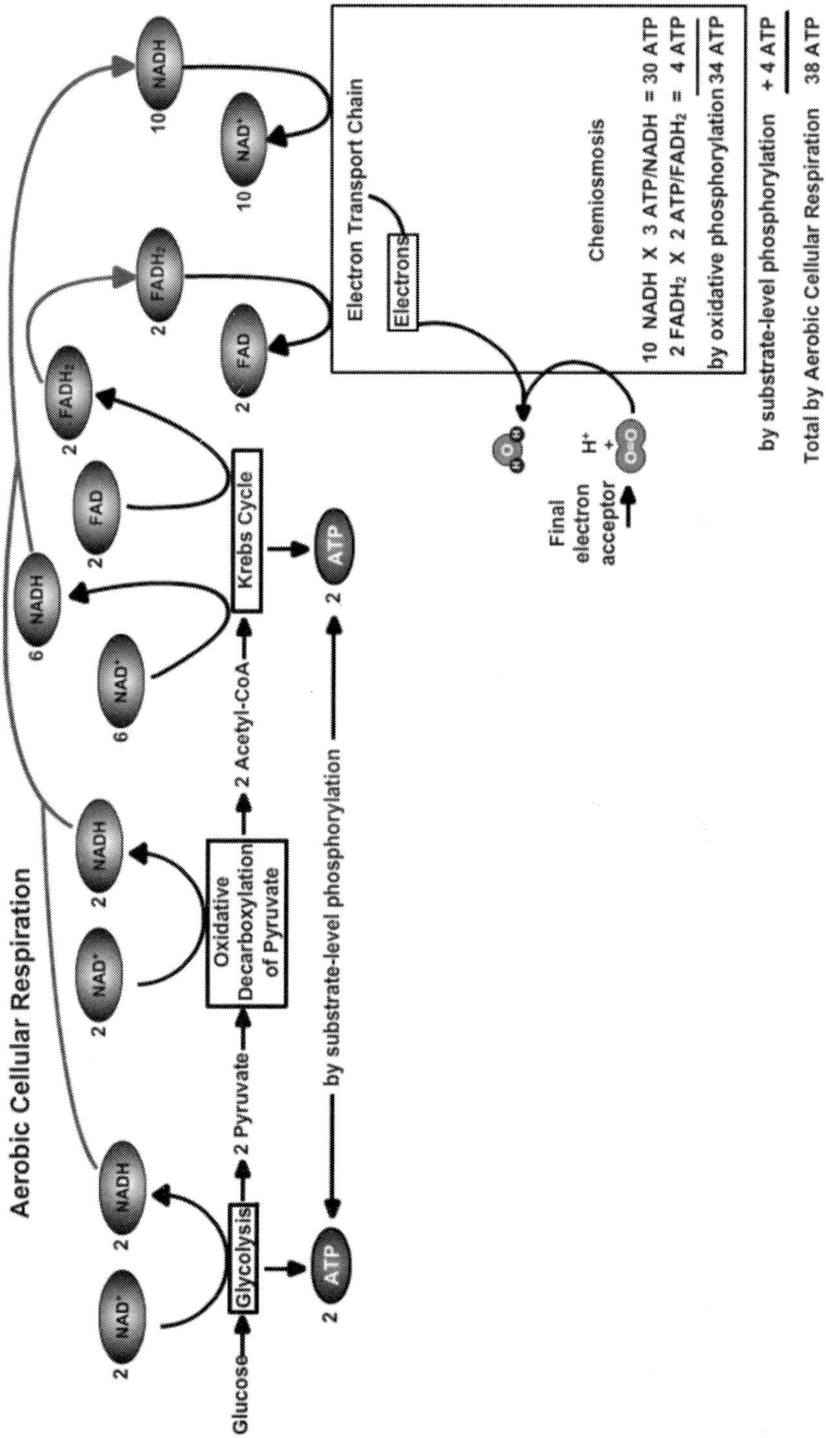

A summary of catabolism in aerobic microorganisms

Energy-Producing Process	Growth Conditions	Final Hydrogen (Electron) Acceptor	Type of Phosphorylation Used to Generate ATP	ATP Molecules Produced per Glucose Molecule
Aerobic Respiration	Aerobic	Molecular oxygen (O_2)	Substrate-level and oxidative	36(eukaryotes) 38 (prokaryotes)
Anaerobic Respiration	Anaerobic	Usually an inorganic substance (such as NO_3^-, SO_4^{2-}, or CO_3^{2-}) but not molecular oxygen (O_2)	Substrate-level and oxidative	Variable (fewer than 38 but more than 2)
Fermentation	Aerobic or anaerobic	An organic molecule	Substrate-level	2

Comparison of aerobic respiration, anaerobic respiration and fermentation

Microorganisms derive energy not only from the oxidation of inorganic compounds, but also from light energy, which they capture and use to synthesize ATP and reducing powers. It is referred as photophosphorylation. The process by which light energy is trapped and converted to chemical energy is called *photosynthesis*.

Photosynthesis is a type of metabolism in which catabolism and anabolism occur in sequence. As a whole, it is divided into 2 parts.

A. Light reaction

B. Dark reaction

Light reaction is a catabolic reaction in which light energy is trapped and converted to chemical energy and electrons or reducing powers.

Dark reaction is an anabolic process in which the energy is used to reduce or fix CO_2 and synthesize cell constituents (carbohydrates), which is also called CO_2 fixation.

For conversion of light energy to ATP, the bacteria possess *light harvesting pigments*. They are *chlorophyll a, carotenoids, phycobiliproteins* (which are present in cyanobacteria) and *bacteriochlorophyll* (which are present in purple sulphur bacteria).

In bacteria, there are two types of light reactions (conversion of light to ATP) and two types of CO_2 fixation occur.

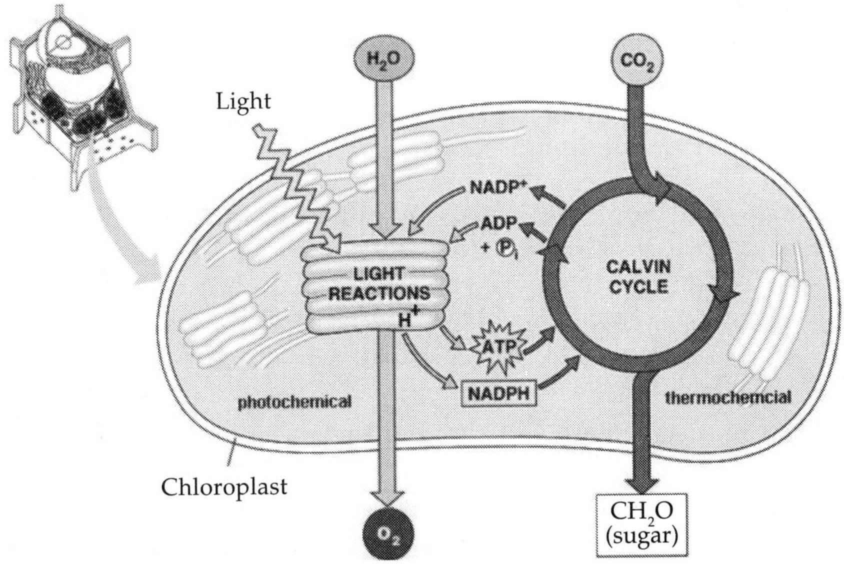

Photosynthesis – Light dependent and light independent reactions

A. Light reaction (Photophosphorylation)

For photophosphorylation, a light harvesting pigment, a membrane electron transport chain, source of electron (electron donor) and ATPase enzymes are required.

Two types of photophosphorylations occur during photosynthesis.

1. Cyclic photophosphorylation
2. Non-cyclic photophosphorylation

- In plant and cyanobacteria, both cyclic and non-cyclic photophosphorylation occurs whereas in purple bacteria, the cyclic photophosphorylation only occurs.
- In plant and cyanobacteria, the electron source is water, by photolysis, H_2O split into H^+ and O_2 and during the process, O_2 is evolved and referred as oxygenic photosynthesis.
- Since, the sulphur bacteria is an anaerobic bacterium, they use H_2S instead of H_2O as electron donor. Since, there won't be any O_2 evolution during photosynthesis, referred as anoxygenic photosynthesis.

1. The oxygenic phosphorylation (Non cyclic photophosphorylation)

In plants, algae and cyanobacteria (oxygenic photosynthetic bacteria), non - cyclic photophosphorylation occurs in photosynthesis. In this process, when a molecule in pigment system II (one of two systems of light reactions) absorbs light, this energy raises the molecule to an excited state and the molecule releases an electron. This electron is transferred to plastoguinone, to cytochrome b, to cytochrome f and finally to pigment system system I. Photophosphorylation occurs with generation of ATP from ADP and inorganic phosphate in the step between cytochrome b and cytochrome f. When pigment system I absorb light, it releases an electron. This electron is transferred from ferredoxin, to flavoprotein, to $NADP^+$. Phosphorylation occurs again between the release of the electron from pigment system I to ferredoxin. This process differs from cyclic phosphorylation because the electron lost by pigment system II is not cycled back to it. Instead, electrons are replaced by pigment system II by the light – generated breakdown of water called **photolysis.**

The energy released in the step between cytochrome b and cytochrome f is used for *photophosphorylation* – the generation of ATP from ADP and inorganic phosphate.

End products of this reaction are ATP, NADPH and O_2, which is used in dark reaction.

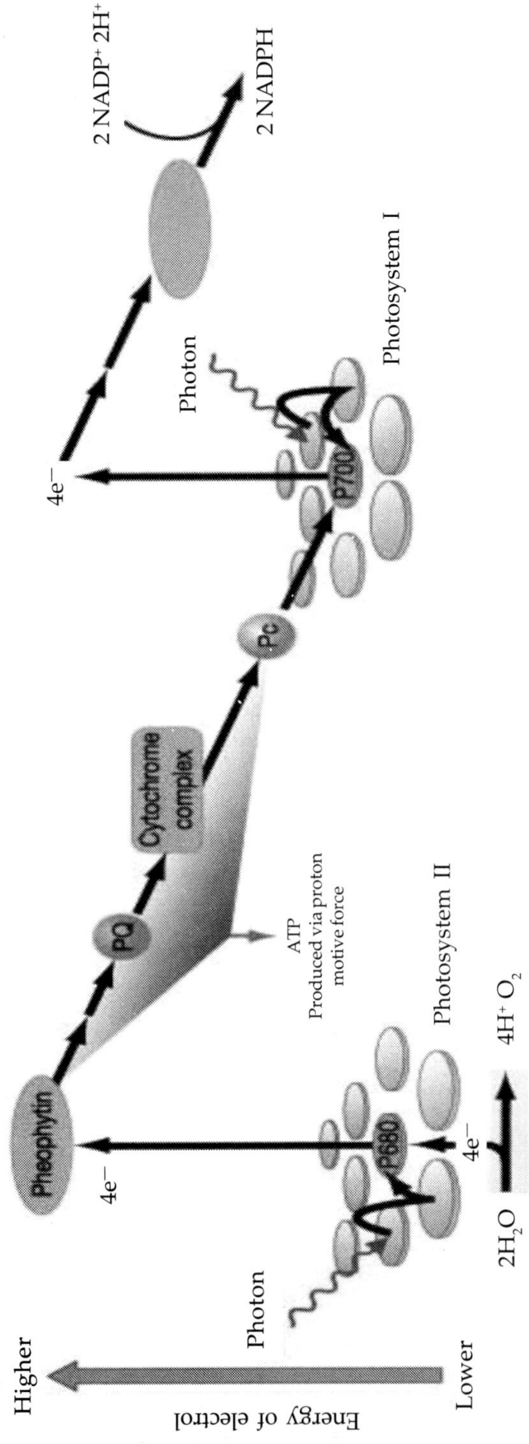

Non cyclic photophosphorylation

2. The anoxygenic photophosphorylation (Cyclic photophosphorylation)

It occurs in photosynthetic bacteria, which posses chlorophylls called bacteriochlorophylls that differ from the chlorophyll of plants in structure and in light absorbing properties. Bacteriochlorophylls absorb light in the infrared region (725 to 1.035 nm). They are not contained in chloroplasts but are found in extensive membrane systems throughout the bacterial cell.

When a molecule of bacteriochlorophyll absorbs a quantum of light, the energy of the light raises the molecule to an excited state. In this excited state an electron is given off by a bacteriochlorophyll. Bacteriochlorophyll thus becomes positively charged. It then serves an electron trap or strong oxidizing agent.

The electron, carrying some of the energy absorbed from light, is transferred to an iron-containing heme protein known as **ferredoxin.** From there it is passed successively to ubiquinone, to cytochrome b, and to cytochrome f and finally back to the positively charged bacteriochlorophyll. Essentially, the electron has gone around in a cycle, beginning with, and returning to, bacteriochlorophyll.

The end products of the reaction are ATP, NADPH and S and these were used for dark reaction.

Difference between plant and bacterial photosynthesis

Organisms	**Plant photosynthesis plants, algae, cyanobacteria**	**Bacterial photosynthesis purple and green bacteria**
Type of chlorophyll	chlorophyll a absorbs 650-750nm	bacteriochlorophyll absorbs 800-1000nm
Photosystem I (**cyclic photophosphorylation**)	present	present
Photosystem II (**noncyclic photophosphorylation**)	present	absent
Produces O_2	Yes (**Oxygenic**)	No (**Anoxygenic**)
Photosynthetic electron donor	H_2O	H2S, other sulfur compounds or certain organic compounds

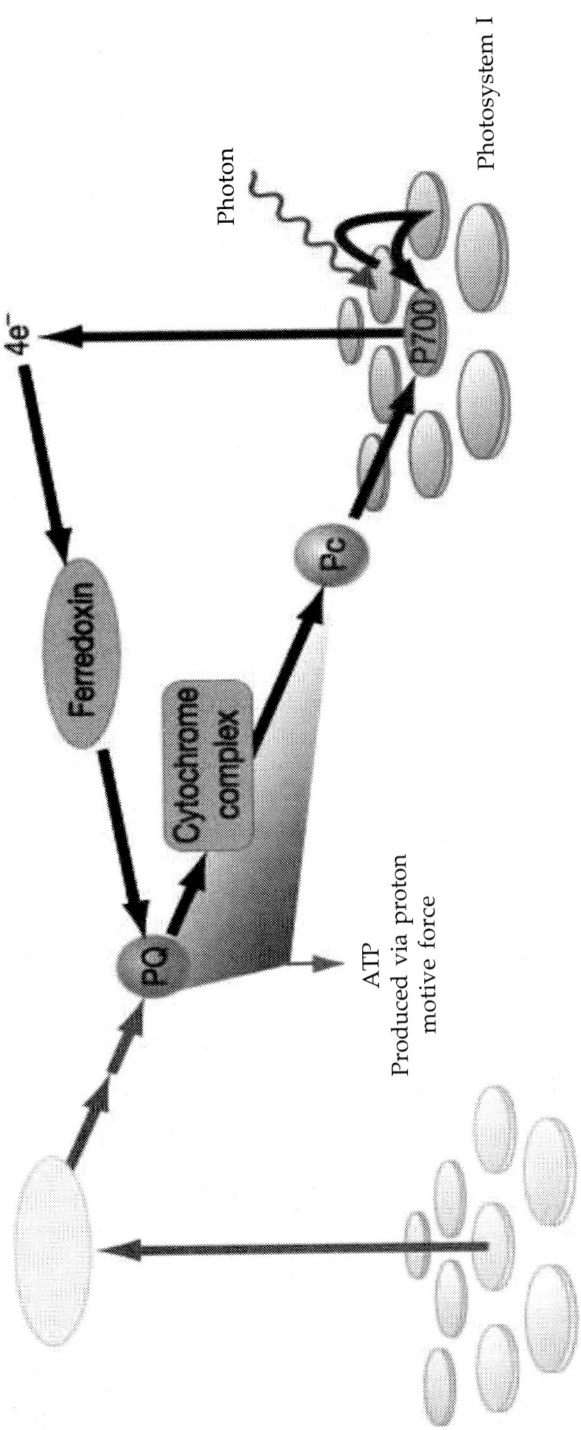

Cyclic photophosphorylation

B. Dark reaction (CO_2 fixation)

The dark reaction in which the ATP and NADPH were used as energy and electron sources to fix the CO_2 as carbohydrates. The pathway involved in the dark reaction is *Calvin cycle*, by which the CO_2 is fixed as phosphoglyceic acid and lead to formation of many sugars. The enzyme RuBiSCO is the key enzyme for this process.

The following pathway shows the Calvin cycle and the formation of key monomers for anabolic reactions such as hexose phosphate – polysaccharides; pyruvic acid – amino acid and fatty acid; pentose phosphate – DNA and RNA.

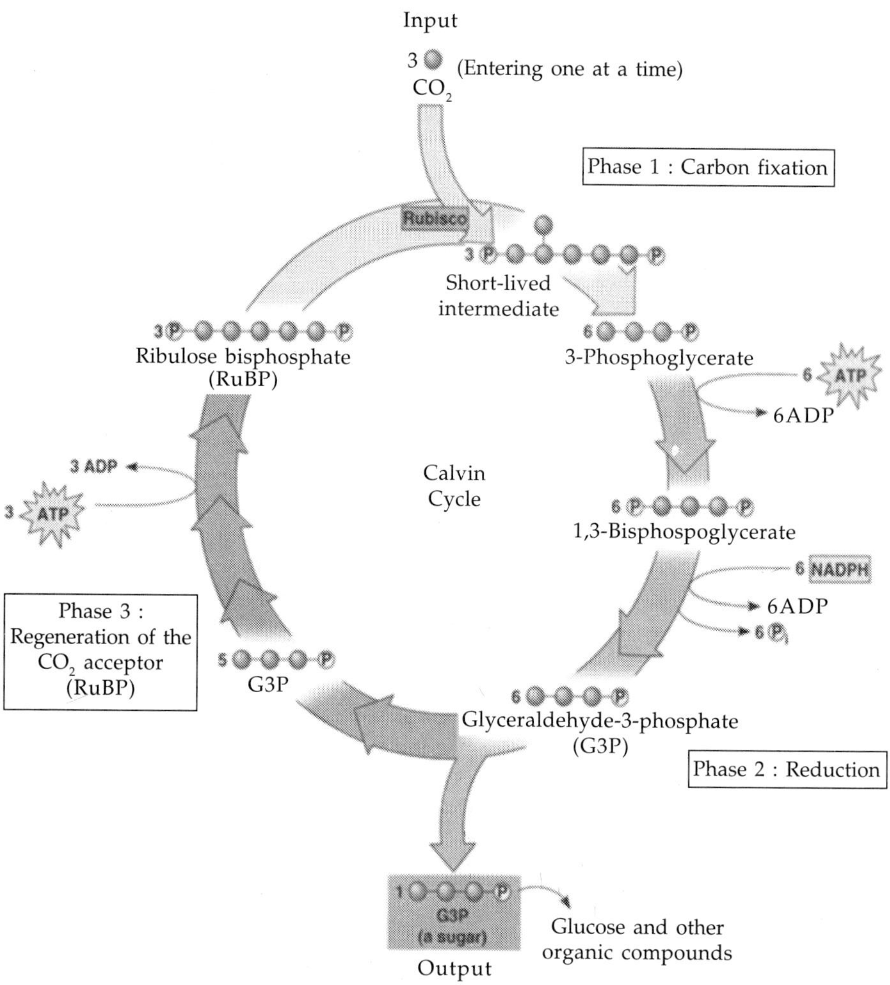

The calvin cycle

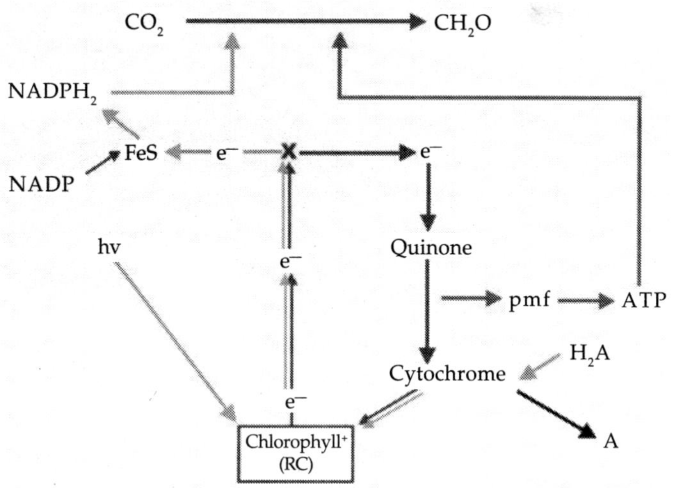

Complete model of light and dark reactions of photosynthesis

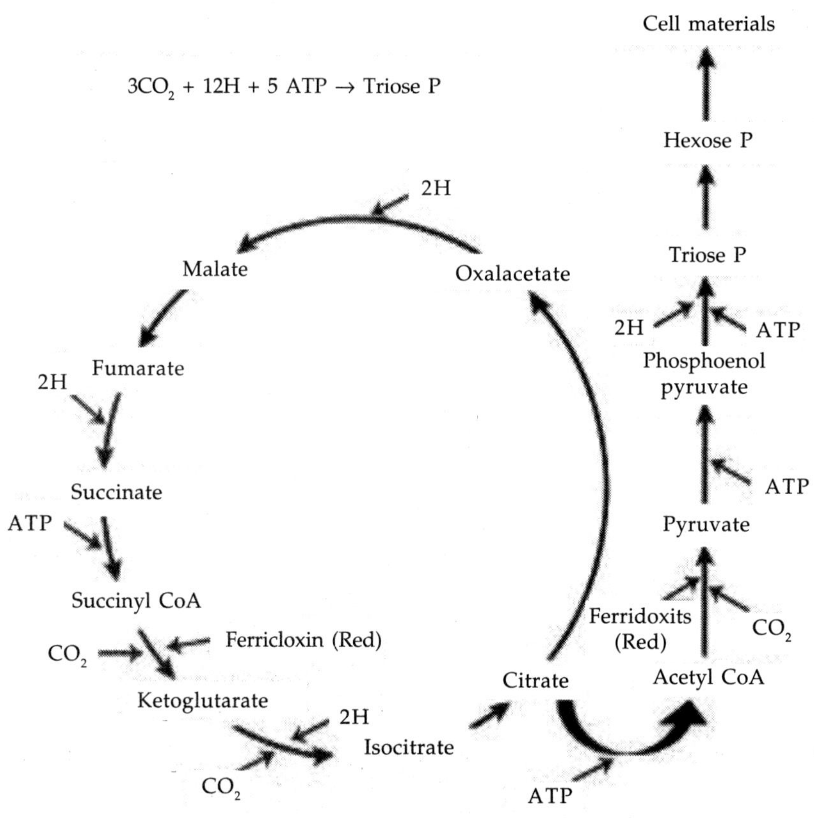

Reverse TCA cycle

Another way of CO_2 fixation by phototrophs

In phototrophs, the electron and energy were derived form sunlight and carbon from CO_2 fixation through Calvin cycle. But some bacteria may derive electron and energy from sunlight and fixes CO_2 by some other path way, not the Calvin cycle. The example is Photosynthetic green bacteria (*Chlorobium*). They derive NADPH and ATP through cyclic phosphorylation, but CO_2 fixation is by *reverse TCA cycle*. Since TCA cycle is *amphibolic pathway* (referring the cycle can operate in both the directions), it can also be used to fix the carbon-di-oxide if operated reversely. The pathway is as follows:

Another way of CO_2 fixation is by methanogens

They use CO_2 as terminal electron acceptor and form CH_4 (methane). They also use acetyl CoA pathway for fixing CO_2.

Organism	**Light reaction/ATP generation**	**Dark reaction/CO_2 fixation**
Cyanobacteria (*Nostoc*), plant and alga	Cyclic and non-cyclic photophosphorylation Oxygenic photosynthesis	calvin cycle
Purple bacteria (*Chromatium*)	Cyclic and non-cyclic photophosphorylation Oxygenic photosynthesis	Calvin cycle
Green bacteria (*Chlorobium*)	Cyclic and non-cyclic photophosphorylation Oxygenic photosynthesis	Reverse TCA cycle

□□□

Part 2

Soil Microbiology

Chapter 12

Introduction and Importance of Soil Microbiology

Soil represents a medium or substrate in which numerous microorganisms live and bring about a great variety of processes which are responsible for continuation of the cycle of life in nature. The numerous living forms which spend all or part of their life in soil may vary in size from sub microscopic forms to the lower animal forms. With the growing recognition of the numerous processes carried out by the microorganisms in the soil there gradually emerged a branch of microbiology, which came to known as soil microbiology. It is a branch of soil science concerned with soil inhabiting microorganisms and their functions and activities.

Recently, soil microbiology has expanded to include the study of the role of soil microorganisms in genetic engineering, in the biological control of pests and diseases, the degradation of pollutants, production and destruction of radioactive gases and its transfer. Thus microbial participation in several important processes emphasizes that soil microbiology has become a global science.

Sergei Winogradsky is considered as the *Father of Soil Microbiology*

Soil microbiology delas with....

- The microscopic organisms of the soil
- Their population and activities
- Role in various transformations taking place in the soil and
- Their importance in plant nutrition and crop production

Distribution of micro organisms in soil

Soil contain five major groups of microorganisms. They are *Bacteria, Actinomycetes, fungi, algae and protozoa*. The soil ecosystem includes these microbial groups as well as the inorganic and organic constituents of a given site. The collections of cells represented in the community are considered as distinct populations. All the inhabitants of the particular locality make up the community.

I. Bacteria

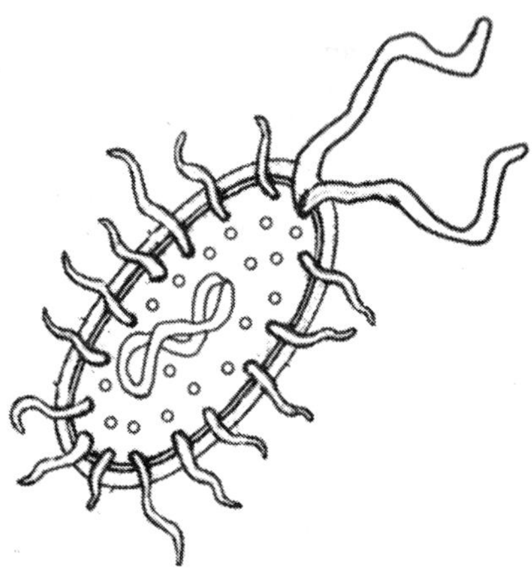

They are the most *dominant group* of microorganism in soil and are more numerous than the other four combined. They present in all types of soil but their population decreases as the depth of soil increases (Horizon A > B > C). The number of cells of bacteria in the soil is always great, but the individuals are small, (μm in length). Because of the minute size of the bacteria and large cells or filaments of the other 4 groups, the bacteria probably account for appreciably *less than half of the total microbial cell mass*. In aerated soil, bacteria, will be dominating, alone are responsible for all the activities in environment lacking O_2 or little O_2. In transformation process bacteria stand first, due to their rapid growth and vigorous decomposition capacity on variety of substrates.

Bergey's manual of Systematic Bacteriology is universally used for the classification of Bacteria. It was discussed in earlier chapters.

Grouping of bacteria

a. Based on Population load

Soil microbiological population has been divided into two broad groups. a. autocthonous and b.zymogenous.

1. *Autochthonous* or native microbes: Indigenous, which are characteristic of the particular soil and which may be expected to be always found in soil. eg.*Arthrobacter*
2. *Zymogenous*, or fermentative organisms require an external source of nutrients and their normal population in soil is low. When specific substrates are added to soil, the population is increased. Then gradually declines when the added substrate is exhausted. Eg. cellulose decomposers, N utilizing bacteria, nitrifiers etc.,
3. *Transient microbes* - comprising organisms that are introduced into the soil.

b. Based on nutrition

1. Autotrophs -synthesize their own food derive energy from light or chemicals.
2. Heterotrophs -depend on preformed food for nutrition, derive energy and C from organic compounds.

Several reactions in nitrogen transformation of soil depend on the chemoautotrophic organisms viz., *Nitrobacter* and *Nitrosomonas* and hence Chemoautotrphy of bacteria in soil is intimately related to crop production.

c. Based on the O_2 requirement

Aerobic - need O_2 for growth

Anaerobic - not require O_2 for growth

Facultative anaerobic - live in the presence or absence of O_2.

d. Based on structure

Depending upon the carbon and energy sources, two sub groups are there in each category such as photoautotrophs, chemoautotrophs, photoheterotrophs and chemoheterotrophs.

Bacilli - Rod shaped

Cocci - Spherical shaped

Spirillum - Spiral shaped

Abundance of Bacteria

Arctic regions are frozen for 9-10 months/year. The bacterial count of 10^6/g is observed even when temperature remains below freezing point. Such organism are in a state of dormancy awaiting for the spring to recover.

Desert soil is another extreme situation. Bacteria exist even in oven dry state. Spore forming bacilli are often predominant.

Importance

- In transformation process bacteria stands first, due to their rapid growth and capacity of vigorous decomposition of variety of substrates.
- Involved in N_2 fixation, P soluilization, CO_2 fixation, S, Fe transformation, Si solubilization etc.
- They are effective molecular vectors.

II. Actinomycetes

A transitional group between the simple bacteria and fungi are the actinomycetes. They are classified under bacteria, but resemble fungi in some aspects. Actinomycetes are sensitive to antibacterial compounds and not antifungal compounds

Characters resembling Fungi

- produces mycelium with extensive branching
- many actinomycetes form aerial mycelium and conidia
- Growth in liquid medium is not turbid as in bacteria, but form clumps and pellets like fungi.

Characters resembling Bacteria

- Some species have flagella that resemble those of true bacteria.
- Similarities in cell wall composition and sensitivity to antibacterial compounds.
- Actinomycetes differ from fungi that they do not have chitin and cellulose which are commonly found in the cell walls of fungi.
- Certain actinomycetes resemble *Mycobacterium* and *Corynebacterium* in all respects both morphologically and physiologically including susceptibility to virus attack.

Common genera in soils

Streptomyces (70%), *Nocardia, Micromonaspora, Frankia* are commonly occuring genera in soil. About 70% actinomycetes are the *Streptomyces.*

Streptomycetes

They elaborate musty odour; an odour reminiscent of freshly turned soil. The metabolite of *Streptomyces* called *'geosmin'* and other volatile products are responsible for the characteristic earthy odor/ smell of soil.

Distribution

Actinomycetes are numerous and widely distributed not only in soil but in a variety of other habitats including composts, river muds and lake bottoms. Present in surface soil and also in lower horizons to considerable depth. In abundance they are second only to the Bacteria and the viable counts almost equal to both. Saprophytic existence, but a few species can cause diseases of plants, domestic animals and even humans.

Abundance

Population is 10^5 to 10^8/g in temperate zone, but lower counts in waterlogged soils, acid peat, arctic, Tundra regions. In alkaline areas, especially when dry, the relative abundance is high.

Importance

- Decomposition of resistant components of plant and animal tissue.
- Carry out the transformations at high temperature particularly in the manure and compost pits – thermophiles are dominant
- Cause certain soil borne diseases of plants, potato scab of apple, sweet potato pox.
- Cause infection to humans, animals.
- Importance in microbial antagonism by production of antibiotics and production of enzymes.
- Have a significant role in antibiotic Industry.

III. Fungi

As the important constituent group of the soil population, they are widely distributed in most well – cultivated soils. Fungi account for a large part of the total microbial population. Though they are not the major inhabitants, they do infact makeup the significant part of the biomass because of the large diameter and extensive network of the filaments. Fungi exist in soil in the form of vegetative mycelium and spores. Characteristically the fungi possess a filamentous mycelium ie., network of individual hyphal stands. The hypha itself is rather broad and has a diameter appreciably greater than that found in the common actinomycetes. In nature conidia or asexual spores are abundant and widespread, the sexual spores are relatively uncommon. In contrast with bacteria they can be effectively differentiated into genera and species on the basis of morphology.

Distribution and Abundance

Estimates of microbial density reveal the presence in soil is ranging from 20,000 to 1,000,000 fungal propagules per gram, the propagule being considered as any spore, or hyphal filament that is capable of giving rise to a colony. The length of the fungal mycelium has been reported to range from 10 to 100m per g surface soil, but sometimes even upto 500-1000 m. Assuming the hyphal filament has an average diameter of 5µm and a specific gravity of 1.2 and taking the range of 10 to 100 m per gram, it would appear that the weight of fungi ranges from 500 to 5000 kg per ha of surface soil. Thus the filaments make up a significant part of the soil biomass.

Individual species and genera have been recorded in diverse and highly dissimilar habitats. They are the inhabitants of peats, flooded soils, regions with low and extremely high salt contents, locations in many deserts, Antartica, as well as in the tundra.

Importance

- Degrade complex molecules-cellulose, hemicellulose, pectin, starch, lignin
- Improve soil structure and helps in aggregate formation
- Express pathogenecity in plants, animals and humans
- Predator against soil protozoa - Hyphal penetration reduces mobility and causes total cessation of the movement
- Trapping nematodes - *Arthrobotrys, Dacylaria, Dactylella, Harposporium*
- Exhibit Beneficial association with plants - Mycorrhizal fungi

Yeast are the important group of fungus and they are non filamentous fungi. Their presence may be demonstrated in most soils. Soil yeast are *Candida, Debaryomyces, Rhodotorula, Torulopsis*. Yeasts have been found in comparable numbers in soils of Antarctica, grasslands, cultivated fields, and forests.

Importance

- Involved in the degradation of complex molecules. They can utilize and degrade the major plant constituents – cellulose, hemicellulose, pectins, starch and lignin.
- Participate in humus formation from fresh organic residues.
- Carry out inorganic transformations and influences the formation of stable aggregates by means of hyphal penetration and the mechanical binding of the particles.
- Pathogenecity is the common character associated with several soil borne fungi. Ex: *Fusarium, Helminthosporium.*
- Causes disease among humans, animals

IV. Algae

They occur in small numbers in soil. They are photosynthetic organisms. Abundant in habitats in which moisture is adequate and light is accessible. Isolation can be done in liquid media and by plating with agar media. Because of their similarities to bacteria, the BGA are classified together with the bacteria.

They follow Photoautotrophic nutrition i.e use light as energy source and CO_2 as C source. In the absence of light heterotrophy occur in some species of *Chlorophyta, Cyanophyta and Diatoms.* These species use the oxidation of organic carbon to replace the light in supplying energy for anabolic process (Called facultative photoautotrophs).

Occurrence of this photo synthetic microorganism has been recorded in soils throughout the world. Algae are moderately adaptable to environmental changes, persist in unfavourable circumstances such as in alkaline and desert soils.

Some species colonise the zone under the surface crusts of limestone and sandstone rocks in the desert, living at sites where the humidity is retained and where sufficient light passes.

- Present as 100 to 50,000/g soil
- Biomass contributes 7 to 300 kg/ha

- Flooded paddy field is the good environment for algae, could have a great agronomic significance. The microbial action may be associated with the utilization of atmospheric N, the release of O_2 or the excretion of products stimulating plant developments. Under water logged soils, an algal film farms at the liquid surface, make up an appreciable mass. *Anabaena, Calothrix, Nostoc, Oscillatoria* are the common inhabitants.

Importance

- Involved in CO_2 fixation
- Contribute to soil formation by colonizing the barren surface, corrode and weather rocks
- Algal layer covers the rocks, on death it favors secondary colonizers
- Surface bloom reduces erosion losses probably by binding together with soil particles
- Improve soil structure, texture and add fertility to soil after decay
- Photosynthesis liberates O_2, provides O_2 to the roots and other microorganisms
- Involved in N_2 fixation
- N_2 is released after decomposition of algal biomass
- Produces several metabolites
- Used in the production of single cell protein (SCP) as well as Beta carotene.

V. Protozoa

The subterranean soil contains protozoa, earthworms, nematodes, insects. Most abundant one is protozoa, the simple form of animal life. Primitive, unicellular and majority are dependent on preformed organic matter. They obtain their nutrition from soluble organic and inorganic substances, or by a phagotrophic nutrition (direct feeding upon microbial cells). They maintain the equilibrium by engulfing.

- Regulate the bacterial community.
- Maintain the biological equilibrium in soil.
- Participate in the decomposition of plant remains.

❑❑❑

Chapter 13

Factors Affecting Microbial Activities in Soil

Several environmental conditions affect the density and composition of the microflora and frequently alter their activities in soil. Primary factors include moisture, aeration, temperature, pH, organic matter and inorganic fertilizers. Lesser variables are the secondary factors which include crop rotation, season, soil depth, cultivation practices etc.

I. Primary Factors

1. Soil moisture

Soil moisture is one of the important factor influencing the microbial population. Water is the major component of protoplasm, an adequate supply must be available for vegetative development. But, when it becomes excessive, proliferation is suppressed due to limitation in gaseous exchange and lowers the availability of O_2 supply creating anaerobic environment. Most of the organisms prefer a moisture percentage between 20 and 60.

Under dry conditions, the bacteria may form spores which can resist the drought conditions. The fungi may sporulate or form chlamydospores to tide over adverse conditions. The protozoans also may form cysts and can survive under dry conditions. Actinomycetes are the chief group of organisms that prefer dry conditions.

At excess moisture, it is believed that the concentration of nutrients is diluted and also the aeration is very much limited and hence only the anaerobic and microaerophilic organisms can develop better.

2. Soil air

It is directly linked to the moisture level of the soil. Most of the forms are active in aerated soil and under waterlogged soils anaerobic and microaerophilic forms develop.

3. Soil temperature

Exhibit considerable influence on the microbial population. Though microorganisms have been found to exist under extreme temperature conditions, such as –60°C and +60°C the soil temperature usually does not reach such extremes. Microbial population varies both quantitatively and qualitatively under extreme conditions. In tropical and subtropical regions, temperatures vary widely in summer and winter and the population may also be varied. In temperate regions, not much variation in temperature is seen during summer and winter and hence not much variation in the soil population. Soil temperature influences the temperature of air, water and solid phase of the soil. Soil water and temperature exert a combined influence on the microbial population.

4. Soil organic matter

Community size is related to the organic matter content, so that humus rich localities have the largest biological numbers. Organic matter content varied with soil types from less than 0.5 per cent in desert soils to 40 per cent in peaty soils. Organic matter being the chief source of energy and food for most soil organisms, it has great influence on the population. Nature of organic matter is responsible for the differential stimulation of the population. There are several indirect effects of the organic matter on soil microflora. It influences the structure and texture of soil, besides enriching with nutrients for plants and microorganisms. Such influences on the soil also greatly affect the activity of the soil microorganisms.

5. Soil pH

It is a key factor influencing the microflora of soil. It influences enzyme system and thus plays an important role in microbial activity.

In general fungi are 3 times better than bacteria and actinomycetes in acid soil condition. Bacteria flourish well in neutral and alkaline soils. The saline and alkaline soils have different microflora. Salinity is due to excess salts, alkalinity is due to high H-ion concentration. Several direct and indirect effects of H-ion and salt concentration in soil are exerted on microbial population.

6. Fertilizer application

Application of fertilizers, to the soil improves the microbial activity because of the availability of more readily obtainable nutrients. Some fertilizers, may however have inhibitory effect on specific bacterial types. Continuous application of ammoniacal fertilizers favours the growth of fungi due to the formation of nitric acid and which inhibits the growth of bacteria and actinomycetes. Addition of nitrate inhibits the activity of free living N_2 fixing bacteria like *Azotobacter*. Some of the autotrophs are encouraged by the addition of fertilizers.

7. Cropping and vegetation

Two kinds of effects are exhibited by the crop plants. One is through root exudates, which may have different compounds with reference to the crops grown. Vegetation have selective stimulation over population. Continuous cultivation leads to more microbial activity than the uncultivated land.

II. Secondary Factors

1. *Crop rotation* : Crop rotation with different species like legumes, graminaceous plants, etc. brings about different stimulatory effect on the microflora. Some crops have deeper roots than others and some are more fibrous. Such variations bring about physical changes in soil which in turn may have direct and indirect stimulatory effects on the soil microflora.
2. *Cultural practices* : Various cultural practices, such as tillage operations and irrigation have several physical and chemical changes in soil which are reflected on the soil microflora. Through subsoil ploughing deeper layers may get better aeration and there may be quicker multiplication of aerobic organisms. Weeding, irrigation etc. may influence the microbial populations in the soil.
3. *Soil depth* : Most of the organisms are almost in top layers, largely in upper few centimeters and decline with greater depth, more

active at few cm depth and less active at deeper layers. Low O_2 and less sunlight in deeper layers reduce population.

4. *Season* : Cell number are greatest during the spring and autumn and a decline occurs during hot, dry and winter seasons. The cells remain in a state of dormancy in biochemical and metabolic activity. The influence is mainly due to the alterations in temperature, moisture as well as availability of organic matter.

III. Specific Influences

- Sunlight favours algae and other autotrophs.
- Herbicide application has devasting effect on algae.
- Attack by neighbours – protozoa, nematode, earthworms consume algae.
- Presence of parasites and predators – eg. viruses eat on specific bacteria
- Increase in bacterial population increases protozoans also.

IV. Other Factors

- *Burning of top soil* : Leads to partial sterilization of the top soil and may kill protozoan population, which may lead to the increase in bacterial population. This condition may affect biological equilibrium in soil.
- *Application of nematicide,* fungicide and bactericide may exhibit partial sterilization. Thus every direct or indirect change in crop production, alters the soil microflora, still many more are not clearly understood.

Chapter 14

Rhizosphere and its Importance

Rhizosphere

The term *Rhizosphere* was introduced by the German scientist Hiltner in 1904. It is the region of the soil which is subjected to the influence of plant roots. It is the unique environment under the influence of plant roots. Rhizosphere is characterized by greater microbiological activity than the soil away from plant roots. This is mainly because of the root exudates which includes aminoacids, organic acids, sugars, vitamins and phytohormones.

The term rhizsophere is often devided into two general areas, the inner rhizosphere (very root surface) and the outer rhizosphere (immediate adjacent soil). The microbial numbers are larger in the inner zone, where the bio chemical interactions between micro organism and roots are most pronounced.

The root surface and its adhering soil are termed *rhizosplane*

Microflora of the root zone

The rhizosphere is a highly favorable habitat for the proliferation and metabolism of numerous microbial types. Microscopic examination reveals the presence of a vast microbial community surrounding on the surface of roots and root hairs.

- Bacteria are quite abundant in the rhizosphere
- Fungi and Actinomycetes are observed but not as abundant as bacteria
- Protozoa are relatively conspicuous, particularly small flagellates and large ciliates.

Rhizosphere effect

It indicates the overall influence of plant roots on soil microorganisms. It can be put on a quantitative basis by the use of R-S ratio. The rate of metabolic activity of the rhizosphere microorganisms is different from those of non-rhizosphere soil since they are highly influenced by the root exudates like aminoacid, organic acid, sugar, vitamins, phytohormones etc.

R:S ratio

It is defined as the ratio of microbial numbers per unit weight of rhizosphere (R) soil, to the numbers per unit weight of the adjacent non rhizosphere soil (S). The rhizosphere effect is consistently greater for bacteria than for other microbe.

$$\text{R: S ratio} = \frac{\text{Microbial numbers per unit weight of rhizosphere soil (R)}}{\text{Microbial numbers per unit weight of the adjacent non rhizosphere soil (S)}}$$

- Short, gram – negative rods, which almost invariably make up the larger % of the rhizophere
- The % incidence of short gram positive rods, coccoid rods and spore forming (bacteria) (*Bacillus* spp.) are lesser.
- No selective stimulation or inhibition of gram variable rods, cocci, non spore – formers etc.
- *Pseudomonas, Flavobacterium, Alcaligenes* are especially common in rhizosphere. Some other species of *Arthrobacter, Brevibacterium, Cornebacterium, Micrococcus, Xanthomonas, Bacillus* are also present.
- Anaerobic (bacteria) are affected by the root, this may be attributed to the reduced O_2 tension resulting from root and microbial respiration.

- Bactèrial density is 10^9/g rhizosphere soil. It may cover 4-10% of the root area
- Fast growers and metabolically active population dominate rhizohsphere. This suggests that the rhizosphere flora has a greater ability to effect rapid biochemical changes than the organisms of fallow land.
- Actinomycetes and algae are not significantly benefited by their proximity to roots, and the R:S ratio may rarely exceed 2 or 3:1.

Factors affecting rhizosphere population

- Proximity of the soil sample to the root- More counts is there in soil closer to the tissue surface.
- Depth of sampling-The frequency of organism declines with depth
- Plant species-Due to variation in rooting habitats, tissue composition and excretion products, legumes encounter a more pronounced rhizosphere effect than grasses and grain crops. Individual plant species exert influence on specific flora
- Duration of the crop-Biennials exert more prolonged stimulation than annuals because of their long growth period.
- The age of the plant – stimulation is detectable in very young seedlings.
- Soil treatment – have little influence on the total number of organism.
- Different plant species in the same field have widely divergent numbers of organism in their rhizosphere than the same species (i.e.) the character of vegetation is more important than the fertility level of the soil.
- Composition of the root exudates - determine to a large extent the microbial composition of the rhizosphere environment.

Root exudates

One of the most important factors responsible for rhizosphere effect is the great variety of organic substances available at the root region by the way of exudates from roots, which directly or indirectly influence the quality and quantity of the microorganism in the root region.

The substances exuded by plant roots include amino acids, sugars, organic acids, vitamin, nucleotides, phytohormones and many other unidentified substances.

Influence of the rhizosphere microflora on plant growth

The rhizosphere community has either a favourable or a detrimental influence on plant development. Since microflora is so intimately related with the root system, partially covering its surface, any beneficial or toxic substance produced by microorganisms can cause an immediate response.

Positive effects

- Improve the soil structure
- Improve soil aeration
- Supply plants with available form of nutrients by mineralization
- Highly involved in nutrient transformations like N fixation, P solubilisation, P mobilization etc
- Provide vitamins, hormones, growth regulators to plants
- Protect plants from invasion of pathogens by competing with them/antibiotic production.

Negative effects

- Immobilization of nutrients in the microbial biomass may occur.
- Removal of O_2 and accumulation of CO_2 in rhizosphere environment affects the root activity.
- Domination by pathogenic microorganisms affect plant health.

❐❐❐

Chapter 15

Microbial Interactions in Soil

Microorganisms live in the soil, not in the form of pure culture, but as complex populations. Each particle of soil contains more than one type of organisms. So, microbial ecosystem of soil is the sum of biotic and abiotic components of soil. Many of these organisms depend upon one another for direct and indirect nutrients. Some complete with one another for energy sources and for the elements and components used as nutrients. This results in the formation of numerous associations among the soil microorganism. The composition of the microflora of any habitat is governed by the biological equilibrium created by the associations and interactions of all individuals found in the community. The microorganisms that inhabit the soil exhibit many different types of associations or interactions. Some of the associations are indifferent or neutral, some are beneficial type of interactions and others are detrimental or negative.

I. Neutral Interaction

1. Neutralism

It is a type of neutral association between two microorganisms which behave entirely independent. Each could utilize different nutrients with out producing metabolic end products that are inhibitory to the other. This may be transitory as the condition changes in the environment. Especially depending on the availability of nutrients, the relationship may change.

II. Beneficial / Positive interactions

1. Symbiosis / Mutualism

It is the type of association in which two organisms depend on one another and both are benefited by the relationship. The manner in which benefit derived varies. Few examples are given below.

a. Symbiosis of algae/Blue green Algae and fungi – *Lichens*

   The algae get benefited by the protection from environmental stresses afforded by the hyphae. While the fungi gets carbohydrate from algae. This association is observed in Lichens.

b. Symbiosis of *Rhizobium* and Legume roots- *Legume nodules* *Rhizobium* supply fixed N_2 to legume plants and receive photosynthates in turn

c. Symbiosis of fungi and plant roots - *Mycorrhizae.* The fungal hyphae mobilize nutrients especially P from the far away soil to the plant roots and get photosynthates from the plants in turn

2. Proto co-operation / Syntrophism

If the mutualistic association involves the exchange of nutrients between two species, that phenomenon is called *syntrophism.* This is the obligatory association of two species, for their existence or for their performance of some reaction. Eg: In a medium deficient in nicotinic acid and biotin, neither *Proteus vulgaris* nor *Bacillus polymyxa* will multiply as the former requires nicotinic acid and the latter requires biotin. In mixed culture, in the same medium however both with grow since the partner bacterium synthesizes the missing vitamins.

3. Commensalisms

It is the type of beneficial association, in which only one species derives benefit while the other is unaffected. One partner can attack a substrate and hence not available to the second organism, but the decomposition results in the formation of products utilized by the second. This occurs commonly in soil with respect to degradation of complex molecules like cellulose and lignin.

a. Many fungi able to degrade cellulose and yield glucose and organic acids. This can serve as a source of nutrient for many bacteria and fungi, which are non cellulolytic

b. The lignolytic microbes degrade lignin into simple compounds which are utilized by the non lignolytic microbes.

III. Negative / Harmful / Deleterious Interactions

One species of microorganism retard or inhibit the growth and metabolic activities of another susceptible species of microorganism for want of space, nutrition and protection. It is done in many ways like competition, amensalism, predation and parasitism.

1. Competition

Competition arises mainly for space and nutrition. In such situations the best adapted microbial species will dominate the other species which are depending upon the same limited nutrient substances.

Eg: Competition between strains of *Rhizobium* derived from soil and those applied with legume seeds at the time of sowing. The better competitor enters the root hairs more frequently and it is responsible for a high % of nodules.

2. Amensalism

It is a negative interaction, in which one species is affected and the other species is not. *Antagonism* is a kind of amensalism. One species of microorganism kills or inhibits the growth of another species of microorganism by production of antagonistic substances especially antibiotics, toxins etc. An antibiotic is a substance produced by one organism and in low concentrations it inhibits the growth of another organism.

Eg. Actinomycetes - *Streptomyces, Micromonospora* and *Nocardia.*

Bacteria - *Bacillus, Pseudomonas*

Fungi - *Penicillium, Trichoderma, Aspgerillus, Fusarium*

Anitimicrobial compounds against fungus are present in the soil, which inhibit the germination of fungal spores. This phenomenon is termed as fungistasis.

Cyanide is produced by certain fungus in concentrations toxic to other microorganisms, and algae elaborate fatty acids which exhibit marked antibacterial activity. Some metabolic products like methane, sulfides and other volatile S compounds also have toxic effects on certain groups of microbes. Ex: *Bacillus thuringiensis* toxin to lepidopteran insect.

3. Predation

Direct attack of one organism on another organism. Predation is one of the most dramatic interrelationships among the microorganisms in nature.

Eg. Protozoa are predators of bacteria (Protozoans are a key factor in limiting the size of bacterial populations for maintaining the equilibrium in the microbial community) Myxobacteria and cellular slime molds also affect by feeding directly on them

Bdellovibrio bacteriovorus is a predator on all Gram negative bacteria (discovered by Heinz stolp and Mortimer Starr in 1963).

4. Parasitism

It is a relationship between organisms in which one organism (parasite) lives in or on another organism (host). In this relationship the parasite is completely dependant upon the host for its survival and hence the host is harmed.

Eg. Bacteriophages are parasites on Bacteria. This phages can replicate only within bacterial cell.

Chytrid fungi parasitize algae

❑❑❑

Chapter 16

Nutrient Cycling

The term soil generally refers to the loose material of the earth surface and is the region that supports the plant life. It consists of five major components such as mineral matter, water, air, organic matter and living organisms. The proportion of these components, varies with soil type and other soil conditions. To maintain the level of these components it is essential that they undergo a regular process of recycling. This process of recycling through various transformation is brought about by different microorganism. Plants and microbes require nutrients in simple form whereas animals need them in complex form. These transformations are taking place in soil through the action of microbes. Nutrients are cycled from their elemental status to inorganic compounds, then to organic compounds and back into their elemental status.

Mineralization : organic compounds broken are down to inorganic compounds or elements

Immobilization : inorganic compounds converted into organic compound or as cell constituents.

Carbon Cycle

The most important element in the biological realm and substance that serve as the cornerstone of the cell structure is carbon. It constituents about 40-50% of all living organisms, yet the ultimate source is the CO_2 that exist in a perennially short supply, only 0.03% of the earth's atmosphere, which undergo a cyclic change from an oxidized to reduced state. The carbon cycle revolves about CO_2 ; its fixation and regeneration.

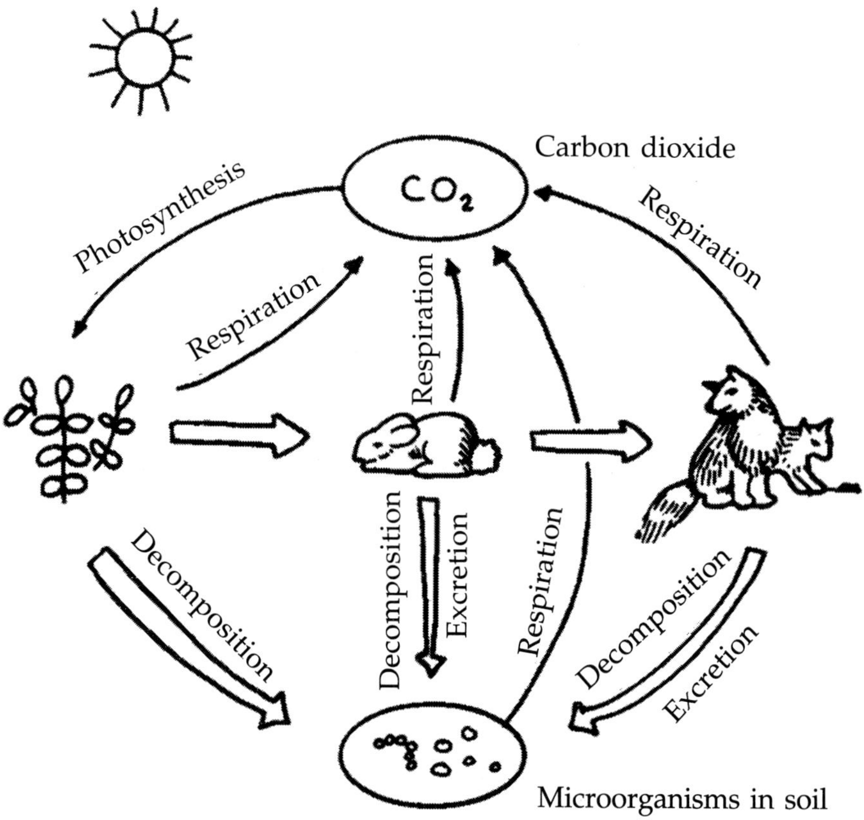

The Carbon cycle

I.Carbon fixation

Carbon (CO_2) is constantly (reduced into organic carbon compounds) being fixed into organic form by photosynthetic organisms (photosynthesis). The green plants utilize CO_2 as their sole carbon source, and the carbonaceous matter synthesized serves to supply carbon to other heterotrophic organisms and animals. It is estimated that $1.3x10^{14}$ kg CO_2 is fixed annually in the biosphere. To the lesser extent CO_2 is also fixed through the agency of photosynthetic bacteria. The chemoautotrophs utilize CO_2 as their sole carbon source. With the conversion of so much of the plant available carbon to organic form each year and the limited supply in the air, it is apparent that the major plant nutrient element would become exhausted in the absence of microbial transformation.

ii. Carbon regeneration

Upon the death of plants and animals, microbes assume a dominant role in carbon cycle. The dead tissues are degraded and transformed into microbial cells and humus or soil organic fraction. Further decomposition of these materials leads to the production of CO_2 and once again it is recycled. Apart from this, the animals and plants release CO_2 through respiration.

The CO_2 thus regenerated is again fixed and this cycle continues.

Soil Organic Matter

The organic matter subjected to microbial decay in soil comes from several sources like Plant remains, forest litter and incorporation of plant tissues, animal tissues and excretory products. Animals and plant residues are made up of complex carbohydrates, simple sugars, starch, cellulose, hemicellulose, pectins, gums, mucilage, proteins fats, oils, waxes, resins, alcohols, aldehydes , ketones, organic acids, lignin, phenols, tannins, hydrocarbons, alkaloids, pigments etc. The chemistry of organic matter is very complex, and each constituent undergoes different transformations by different microorganisms.

Organic matter is a natural substrate for saprophytic micro organism and provides nutrition to plants indirectly through the activity of soil microorganisms. It is essential for the formation of soil aggregates and hence soil structure which ultimately determines the soil aeration and rooting habit of plants. Organic matter helps in the conservation of soil nutrients by preventing erosion and surface run off of nutrients.

The organic constituents of the plants are commonly divided into six categories.

a) Cellulose - Most abundant, constitutes 15-60% of the dry weight of plants
b) Hemicellulose - 10-30% of the plant dry weight is occupied

c) Lignin - 5-30% of the plant dry weight
d) Water soluble fraction- 5-30%, included simple sugar, amino acids
e) Ether and alcohol soluble constituents - a fraction containing fat, oils, waxes, resins and a number of pigments
f) Proteins - As the plant ages, the content of water soluble constituents, proteins and minerals decreases and the % of abundance of cellulose, hemicellulose and lignin increases.

Organic Matter Decomposition

- The soil microorganism play important role in the decomposition of soil organic matter.
- Bacteria are the dominant group – mostly heterotrophic organisms (use energy from organic sources such as sugars, starch, cellulose and protein) – are involved. Autotrophic organism which occupy a small portion of the biomass in soil (use inorganic sources such as Fe and S) are not directly involved in organic matter decomposition.
- Actinomycetes grow on complex substances such as keratin, chitin and other complex polysaccharides and play active role in humus formation.
- Soil fungi are mostly heterotrophos and use organic residues easily
- Soil algae contribute a small amount of organic matter through their biomass, but they do not have any active role in organic matter decomposition.

Organic matter decomposition serves two important functions

a) Provide energy for growth
b) Supply carbon for the formation of new cell materials

Hence only heterotrophs are actively involved in the process of decomposition. The relationship between organic matter and plant growth may be direct or indirect. During the decomposition of organic matter three separate simultaneous processes can be distinguished. The important changes during decomposition are:

1. Plant and animal tissues constituents disappear under the influence of enzymes

2. Synthesis of new microbial cells so that proteins, polysaccharides and nucleic acids typical of bacteria and fungi appear.

3. Third, certain end products of the breakdown are excreted into surroundings there to accumulate or to be further metabolized.

Carbon Assimilation

The process of converting substrate to protoplasmic carbon is known as assimilation. Under aerobic conditions 20-40% of the substrate carbon is assimilated, the remainder is released as CO_2. Fungi are more efficient, in their metabolism, since they convert carbon into cell carbon as filaments and release less of CO_2. 30-40% of substrate C is used to form new mycelium during the decomposition. Compared to fungi, bacteira are less efficient. Aerobic bacteria are less efficient than anaerobic bacteria.

C Mineralization

Conversion of organic carbon substance into inorganic form of carbon which are in available form to the plants.

Immobilization

Assimilation of nutrients by micro organism and through mechanism they reduce the quantity of plant available nutrient in soil. Mineralization is considered better than immobilization.

Importance of organic matter decomposition

1. Important function is the breakdown of organic matter by which CO_2 available for photosynthesis is replenished

2. Any compound that is synthesized biologically is subject to destruction by soil inhabitants, otherwise the compounds would have accumulated in vast amounts on the earth's surface

3. Since, organic matter degradation is a property of all heterotrophs, it is commonly used to indicate the level of microbial activity.

Methods to evaluate the decomposition rate

a) Measurement of CO_2 evolution or O_2 uptake

b) Determination of decrease in organic matter either chemically or by weight loss

c) Observations on disappearance of specific constituents such as cellulose, hemicellulose or lignin.

Changes during organic matter decomposition

As a result of development of mixed flora on chemically complex natural products, some components quickly disappear while others are less susceptible to microbial enzymes persist. The water soluble fraction contains the least resistant plant components and is thus the first to be metabolized. Cellulose and hemicellulose on the other hand disappear not as quickly as water soluble substances, but their persistence usually is not too great. The lignins are highly resistant and consequently become relatively more abundant in the residual, decaying organic matter. In aerobic conditions when carbonaceous substrates are incorporated into soil, immediate drop in O_2 and an increase in CO_2 content of soil air occurs. Change in oxidation reduction potential (En) – it is shifted to a more reduced condition (fall in oxidation reduction potential) occur.

The end products of decomposition are

CO_2, H_2O NO_3, SO_4, CH_4, NH_4 and H_2S may be formed depending on the availability of air.

Factors influencing the organic matter decomposition

- Organic matter level of the soil
- Cultivation
- Temperature
- Moisture
- pH
- Depth
- Aeration
- Nature and abundance of micro organisms involved.
- The extent of availability of C, N, P and K
- Presence of inhibitory substance.

Importance of C:N ratio in organic matter decomposition

Nitrogen is a key nutrient substance for microbial growth. If N content of the substrate is high it is readily utilized and decomposition is faster. If N is poor, decomposition is slower, needs additional N. Protein rich substrates are readily decomposed. Low N or wide C:N ratio results slow decay. Optimum level of C: N ratio for maximum decomposition is 20-25 (1.4-1.7% N). Less than this range, more microbial cells, faster mineralization and more immobilization of nutrients. Wider the ratio, lesser microbial cells, slower the immobilization and mineralization increases gradually, resulting in accumulation of Ammonia and Nitrates. Microbes scavenge the soil solution to obtain enough N. At optimum level, there must be an equilibrium between Mineralization and Immobilization. Soil N level constrains the maintenance of C:N (organic carbon /soil organic matter).

Anaerobic decay / Decomposition of organic matter

The main products of aerobic carbon mineralization are CO_2, water, cells and humus components. In the absence of O_2 organic carbon is incompletely metabolized, intermediary substances accumulate, and abundant quantities of CH_4 and smaller amounts of H_2 are evolved. Energy yield during anaerobic fermentation is low, resulting in fewer microbial cells per unit of organic carbon degraded. Consequently, organic matter breakdown is consistently slower under total anaerobiosis than in environments containing adequate O_2. The rate in water logged soils is intermediate between the two extremes.

When a soil is water logged or flooded there is a shift from aerobic to anaerobic decomposition. Formation and accumulation of organic acids *viz.*, acetic, formic, butyric, lactic and succinic acids appear too, these are frequently detrimental to root development. Organic acids accumulate because of the fermentative character of the microflora of wet soils. The anaerobic carbon transformation are thus characterized by the formation of organic acids, alcohols, CH_4 and CO_2 as major end products.

Under anaerobic conditions, decomposition of organic residues takes place by the activity of both mesophilic, thermophilic microorganisms resulting in the production of CO_2, H_2, ethyl alcohol, and organic acids. Among mesophiles, bacteria are more active than fungi or actinomycetes in cellulolytic activity. They belong to genus *Clostridium* which are numerous in manure pits but rarely encountered in cultivated aerable soils. In compost pits both meso and thermophilic (bacterial and actinomycetes) are important in the breakdown of cellulose substances.

The primary microbial colonizers, initially breakdown the complex carbohydrates and proteins into organic acids and alcohols. At a later stage, the methane bacteria which are strict anaerobes begin to act upon the secondary substrate chiefly lactic and butyric acids and ferment them into CH_4 and CO_2.

Humus

The organic fraction of soil is often termed humus. It is a product of synthetic and decomposing activities of the microflora. Since it contains the organic C and N needed for microbial development, it is the dominant food reservoir. Because humus is both a product of microbial metabolism and an important food source, the organic fraction is of special interest.

Characters of humus

- Humus is the end product of soil organic matter decomposition
- It is a dark coloured and fairly stable soil organic matter with known and unknown physical and chemical properties
- It is an integral part of the organic matter complex in soil
- Humus can be defined as ligno protein complex containing approximately

  45% lignin compounds

  35% amino acids

  11% carbohydrates

  4% cellulose

  7% Hemicellulose

  3% fats, wax, resins

  6% other miscellaneous substances, including plant growth substances and inhibitors.
- Age and composition of the humus are dependent on its origin and environment.
- Bacteria and algal protoplasm contribute a good deal to the nutritive value of humus
- Soilmicro organism take part in humus formation. Some fungi such as *Penicillium, Aspergillus* and actinomycetes produce dark humus like substances which serve as structural units for the synthesis of humic substances.

Benefits of humic substances

- Improve seed germination, root growth, uptake of minerals by plants and other physiological effects on plant growth
- Increases the enzyme activities involved in plant metabolism since humic acid serves as hydrogen acceptors.
- Increases the cytochrome oxidase activity in root systems results in growth stimulatory effect (on roots)
- Chelating effect – on trace elements Fe uptake by roots
- Enhances vigour and yield of plants
- Humic acid known to influence the growth and proliferation of microorganism
- The numbers of *Aspergillus niger*, *Penicillium*, *Bacillus* sp., *Azotobacter* are enhanced

Humus formation

- Once the plant or animal remains fall on or are incorporated into the soil, they are subjected to decomposition
- From the original residues, a variety of products are formed
- As the original materials and the initial products undergo further decomposition, they are converted to brown or black organic complexes
- At this stage any trace of the original remains no longer remains
- The native organic fraction originates from two sources: the original plant debris entering the soil and the microorganism with in the soil body. The microorganism in soil body work upon the former (i.e. plant debris) and synthesize microbial protoplasm and new compounds that become part of the organic fraction.
- Humus exist in a dynamic state
- Chemistry of humus is complex
- It has been pointed out that the organic fraction is derived from

a) Plant constituents that are modified by the microflora.

b) Constituents of microbial cells and products of microbial metabolism that are relatively resistant to decay and therefore persist for sometime after death of organism.

Composition of Humus

The organic fraction contains compounds of C, H, O, N, P, S and small amounts of other elements. Only a small portion of the total is soluble in water, but much can be brought into solution by alkali.

No definite composition can be assigned. It is considered as a portion of the soil that is composed of a heterogenous group of substances, most having an unknown parentage and an unknown chemical structure. Humus contains a number of polymerized substances, aromatic, molecules, polysaccharides, ascorbic acids, polymers of uronic acids and P containing compounds. Lignin and lignin derived molecules have long been considered to be of significance in the formation of humus.

It is possible either that simple aromatics released in the microbial attack on lignin polymerize to yield constituents of the soil organic fraction or that partially altered lignin itself give rise to humus constituents. The monomeric portions of humus are similar to the constituents of lignin.

Degradation processes

(1) Cellulose

Cellulose is a carbohydrate composed of glucose units bound by β-linkage at carbon 1 and 4 of the sugar molecule. It is a polymer of glucose. Woody materials have more cellulose and succulent tissues have poor, but the concentration increases as the plant matures. Cellulose breakdown in soil is influenced by several environmental factors.

Aerobic organism convert cellulose to 2 major products : CO_2 + cell substance, but certain group releases small amount of organic acids. It is however resistant to microbial decomposition. When cellulose is associated with pentosans (xylan, mannans) it undergoes rapid decomposition. When associated with lignin, the decomposition rate is very low. Degradation is by the enzymes that converts cellulose into glucose. (Exoenzymes)

Exoglucanase

Endoglucanase

B-glucosidase (cellulse complex)

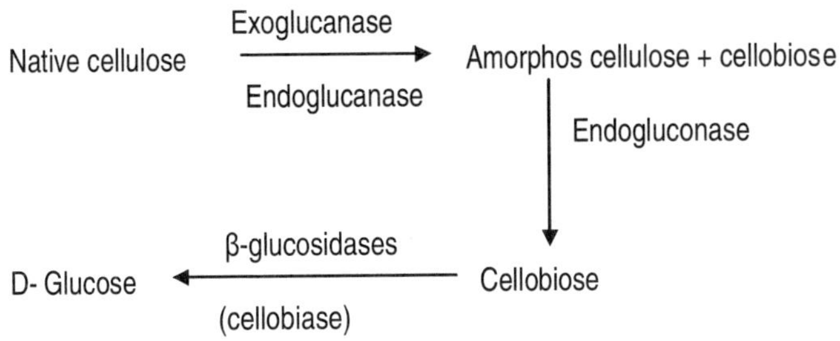

Most cellulolytic bacteria do not excrete significant amounts of cellulase but fungi are found to excrete these enzymes. The soluble sugars released by enzymatic hydrolysis are later utilized by the same or other micro organism for biosynthetic purpose.

(2) Hemicellulose

Polymer of simple sugars such as pentose, hexose and uronic acids. May be either homo or hetero polymers. When they are added to soil, degradation takes place at faster rate in initial stages. The hemicelluloses such as mannans are decomposed rapidly while galactans (polymer of galactose) are decomposed slowly. Many soil microorganisms utilize hemicellulose in both aerobic as well as anaerobic conditions. The microbial degradation occurs through the agency of extracellular enzymes called hemicellulases.

(3) Lignin

Third most abundant costituent of plants. It consists of heterocyclic aromatic organic molecules containing C, H and O. The degradation is very slow and rate of decomposition depends on the presence of other compounds such as cellulose and hemicellulose. Lignin is highly resistant to microbial degradation. Degradation is a complex process.

Upon degradation,

Lignin ⟶ coniferyl ether ⟶ coniferyl alcohol ⟶ coniferyl aldehyde ⟶ vanillin ⟶ vannillic acid ⟶ protocatechuic acid ⟶ ring cleavage

Genera of microorganisms capable of utilizing different components of organic matter as reported by several workers: (F-Fungi; B-Bacteria; A-Actinomycetes)

Nature of substrate in organic matter		Genera of microorganisms
Cellulose	F	*Alternaria, Aspergillus, Chaetomium, Fomes, Fusarium, Rhizoctonia, Rhizopus, Trichoderma, Trichothecium*
	B	*Achrombacter, Bacillus, Cellulomonas, Clostrium, Cytophaga, Pseudomonas, Sporocytophaga, Vibrio*
	A	*Micromonospora, Nocardia, Streptomyces, Streptosporangium*
Hemicellulose	F	*Alternaria, Fusarium, Trichothecium, Aspergillus, Rhizopous, Chateomium, Helminthosporium, Penicillium, Fomes, Polyporus*
	B	*Bacillus, Achromobacter, Pseudomonas, Cytophaga, Sporocytophaga, Lactobacillus, Vibrio*
	A	*Streptomyces*
Lignin	F	*Clavaria, Arthrobotrys, Cephalosporium, Humicola*
	B	*Pseudomonas, Flavobacterium*

□□□

Chapter 17

Nitrogen Cycle

Biological availability of N, P and K is of considerable economic importance, since they are the major plant nutrients derived from the soil. Of the three, N stands out as the most susceptible one to microbial transformations. This element is the key building block of the protein molecule upon which all life is based, it is an indispensable component of the protoplasm of plants, animals and micro organism.

Molecular N_2 constitutes about 78% of the earth's atmosphere but it is chemically inert and cannot be utilized by more living organism, plant animals and micro organism therefore depend on a source of combined N such as ammonia, nitrate or organic N compounds for their growth.

Nitrogen undergoes a number of transformations that involve organic, inorganic and volatile forms of nitrogen. A small part of the large reservoir of N_2 in the atmosphere is converted to organic compounds by certain free living micro organism or by plant microbe association, that makes the element available to plant growth. The nitrogen present in the proteins or nucleic acids of plant tissue is used by animals. In the animal body, the N is converted to other simple and complex compounds. Upon the death, plants and animals undergo microbial decay and organic N is released as ammonium, which is then utilized by vegetation or is oxidized to nitrate by microorganisms. The nitrate from of N is mostly used by the plants or may be lost by bacteria reduced to gaseous N_2, which escapes to atmosphere, there by completing the cycle. The Nitrogen cycle mainly includes transformations such as

1. *Nitrogen mineralization* : In which N containing organic complexes are decomposed and converted into inorganic compounds for use by plants
2. *N immobilization* : In which N containing inorganic compounds are assimilated

N_2 is acted on by certain micro organism sometimes in symbiosis with a higher plant, which can use it as a N source for growth. This process of nitrogen fixation, results in the accumulation of new organic compounds in the cells of responsible micro organisms. The N_2 thus fixed re enters general circulation when the newly formed cells are inturn mineralized.

By means of these reactions the subterranean microflora regulates the supply and governs the availability and chemical nature of N in soil.

Nitrogen Cycle

I. Nitrogen mineralization

The conversion of organic N to the more mobile, inorganic state is known as nitrogen mineralization. As a consequence of mineralization, ammonium and nitrate are generated and organic N disappears. This takes place in two distinct microbiological steps.

(1) Ammonification

It is the process of mineralization in which proteins, nucleic acids and other organic components are degraded by micro organism with the eventual liberation of ammonia. This is called ammonification. A part of the liberated ammonia is assimilated by the micro organisms themselves. The first step in ammonication process is the hydrolysis of proteins, nucleic acids and other organic nitrogenous compounds into amino acids (proteolysis). The amino compounds are then deaminated to yield ammonia. Ammonification usually occurs under aerobic conditions while under anerobic conditions protein decomposition leads to conversion of ammonia into amines and related compounds (eg) *Clostridium* sp. The anaerobic decomposition of protein called as putrefaction. These amines are subsequently oxidized in the presence of O_2 to release ammonia.

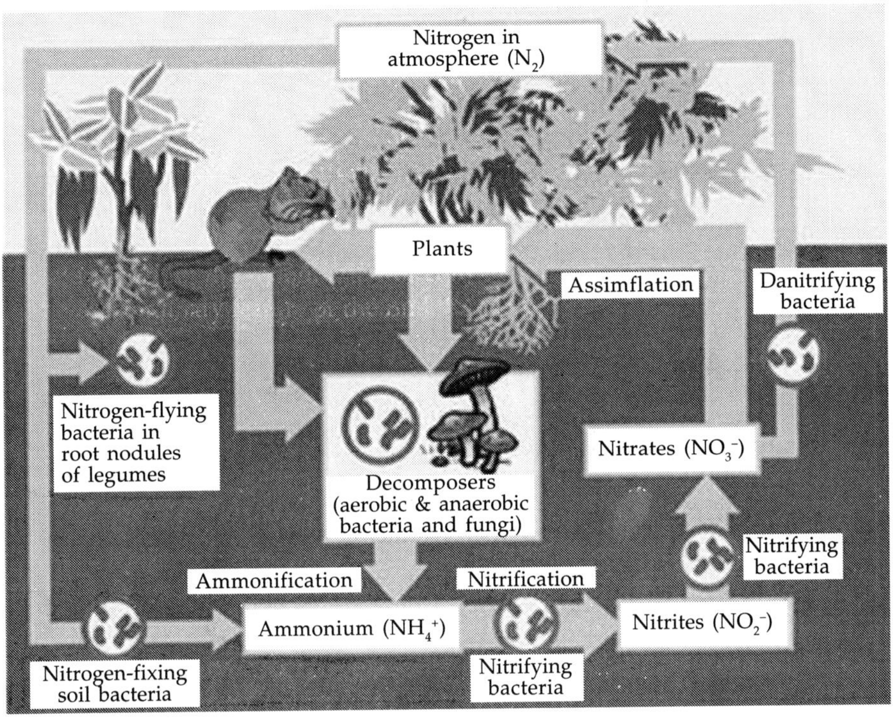

Break down of nitrogenous substance is brought about by the activity of a multitude of microbial species.

Almost all bacteria, actinomycetes and fungi can bring about proteolysis and the amino acids produced are utilized for the growth of these organisms.

Under aerobic condition

Protein/ Nucleic acid/

$$\text{Organic N} \xrightarrow[\text{Proteolysis}]{} \text{Amino acids} \xrightarrow[\text{Deamination}]{} \text{Ammonia}$$

Under Anaerobic condition

$$\text{Protein/ Nucleic acid/Organic N} \longrightarrow \text{Amines}$$

$$\text{(Aerobic) Amines} + O_2 \longrightarrow \text{Ammonia}$$

(2) Nitrification

The biological oxidation of ammonium salts (in soil) to nitrites and the subsequent oxidation of nitrites to nitrates is called as nitrification. i.e. the biological convention of N in soil from a reduced to a more oxidized state is called nitrification. High levels of ammonia and nitrite, good aeration, optimum moisture and temperature, neutral pH and low organic matter favor the nitrification process. In acid soils – nitrification is poor. Since waterlogged soils are deficient in O_2 they are not congenial for nitrification.

Nitrification occurs in two steps;

First step

First ammonia is oxidized to nitrite.

$2\ NH_3 + 1\frac{1}{2}\ H_2O_2 \rightarrow NO_2^- + 2H + H_2O$-Nitrosofication

This change is brought about by chemoautotrophic bacteria of the genera *Nitrosomonas, Nitrosolobus, Nitrosococus, Nitrosospira*. These bacteria obtain their energy requirement by the oxidation of NH_4^+ to NO^-_2. Among the nitrifiers *Nitrosomonas* are most important in soils. Some heteotrophs like *Streptomyces, Nocardia* are also involved.

Second step

Nitrite is further oxidized to nitrate

$HNO_2 + \frac{1}{2}\ O_2 \rightarrow HNO_3$.

Organisms : *Nitrobacter, Aspergillus, Penicillium, Cephalosporium.*

II. Denitrification

The conversion of nitrate and nitrite into molecular N_2 or nitrous oxide through microbial processes is known as denitrification. Certain bacteria are capable of using nitrate as the terminal electron acceptor under anaerobic conditions. This is called *nitrate respiration*. As a consequence of nitrate respiration, NO_3 is reduced to N_2 gas or nitrous oxide. Denitirifcation leads to the loss of N from the soil. It depletes N, and therefore it is not a desirable reaction. Though it is an undesirable reaction in point of view of plant nutrition, but has ecological importance. With out denitrification, the supply of N on the earth world have got depleted and NO_3 would have accumulated. High concentration of NO_3 are toxic, denitrification is a mechanism by which some of the N is released back to the atmosphere.

The escape of molecular N into the atmosphere is also known as *volatalization*.

Denitirfication occur mostly in waterlogged anaerobic soils with a high organic matter contents. Fallow soils flooded with water are more congenial for denitrification than well drained and continuously cropped soils. Denitrification of bound nitrogen to gaseous N is mediated by numerous species of bacteria, which normally use O_2 as hydrogen acceptor (aerobically) and, also use nitrates and nitrites (anerobically).

Anaerobic conversion of nitrate into molecular nitrogen is known as nitrate respiration.

Steps involved

Nitrate is first reduced to nitrite, which is then transformed to nitrous oxide (NO). The nitrous oxide is converted to N_2 with N_2O as an intermediate.

$$2NO^-_3 + 10\ H \rightarrow N_2 + 4H_2O + 2OH^-\ \text{(or)}$$

$$2NO^-_2 + 6\ H \rightarrow N_2 + 2H_2O + 2OH^-\ \text{(or)}$$

$$N_2O + 2H \rightarrow N_2 + H_2O$$

The enzymes involved

$$2\ HNO_3 \xrightarrow{1} 2HNO_2 \xrightarrow{2} 2\ NO \xrightarrow{3} N_2O \xrightarrow{4} N_2$$

1. Nitrate reductase
2. Nitrite reductase
3. Nitric oxide reductase
4. Nitrous oxide reductase

Since nitrates are used as a source of electron acceptor, there is a net loss of N from soil. This process is also termed as *dissimilatory nitrate reduction*. Many soil bacteria like *Thiobacillus denitrificans* oxidize S (chemoautotrophically) and also reduce nitrate to nitrogen.

$$5S + 6\ KNO_3 + 2\ H_2O \rightarrow 3N_2 + K_2SO_4 + 4KHSO_4\ \text{(or)}$$

$$5\ K_2S_2O_3 + 8\ KNO_3 + H_2O \rightarrow 4N_2 + 9\ K_2SO_4 + H_2SO_4$$

Organisms: *Pseudomonas, Achromobacter, Bacillus, Micrococcus*

III. Nitrate reduction

The reverse of nitrification process. That is the reduction of nitrate to nitrite and then forms ammonia.

$$HNO_3 + 4H_2 \rightarrow NH_3 + 3H_2O$$

Since organisms are able to obtain cellular N through ammonia assimilation, the process is called as assimilatory nitrate reduction.

IV. Nitrogen immobilization

The process of microbial assimilation of inorganic nitrogen is referred as immobilization. In contrast to mineralization microbial immobilization leads to the biosynthesis of the complex molecules of microbial protoplasm from ammonium and nitrate. Immobilization results in a marked depression of nitrogen uptake by the plant.

The mineralization of organic N and the microbial assimilation of inorganic ions proceed simultaneously. Both mineralization and immobilization takes place regardless of the % of N in the organic N in organic matter. On the death of micro organism, the immobilized N is however released through mineralization. It is also a loss of nitrogen. NO_3 when accumulated in microbial protoplasm it is referred as assimilatory NO_3 reduction.

V. Biological N_2 fixation (BNF)

Fixation of elemental nitrogen in the atmosphere by the micro organism through a reductive process into ammonia is called as BNF. A variety of prokaryotic organism have the ability to reduce the atmosphere N_2. BNF accounts for about 70% of the total N fixed in the biosphere. The ability to reduce atmosphere N is restricted only to bacteria, which are belonging to the diverse groups. The root nodule associations were the first to be recognized for their ability to fix atmosphere N_2. Rhizobia are the first group of organism realized for its potential of nitrogen fixation.

Nitrogen fixing bacteria

Nitrogen fixing bacteria are classified according to their mode of fixation.

a. Free living nitrogen fixers

   Azotobacter - Fixes under aerobic condition

   Beijerinckia – Prefers acid condition, and aerobic condition

Clostridium – Anaerobic Na fixer

Cyanobacteria (Blue green algae), etc. - aerobic Na fixer

Species of Azotobacter

A. chroococcum

A. vinelandii

A. beijerinkii

A. paspali

A. agilis

A. insignis

A. macrocytogens

b. Associative symbiotic nitrogen fixer

Azospirillum

Herbaspirillum

Species of *Azospirillum*

A. lipoferum

A. brasilense

A. amazonense

A. halopareferans

A. irkense

A.largimobilis

c. Endophytic nitrogen fixer

Gluconacetobacter diazotrophicus

d. Symbiotic nitrogen fixers

Rhizobium (*Rhizobium – legume* association)

Bradyrhizobium (*Bradyrhizobium – soybean* association)

Azorhizobium (*Azorhizobium- Sesbania rostrata* association)

Anabaena azollae (*Azolla – Anabaena* association)

Frankia (*Frankia – Casuarina* association)

Important genera of blue green algae

Anabaena, Nostoc, Cylindrospermum, Rivularia, Oscillatoria, Plectonema, Aphanothece, Lyngbya, Scytonema, Calotrhix etc.

Species of Azolla : *A. pinnata, A. filiculoides, A. microphylla, A. caroliniana, A. mexicana, A. nilotica*

Rhizobium

Rhizobium is a predominant symbiotic N_2 fixing bacterium. Boussingault showed that leguminous plant can fix atmospheric N_2. Then hellriegel and wilfarth – proved that N_2 is fixed by certain bacteria living in root nodules of leguminous plants. Latter isolated in pure culture by *Beijerinck*. Winogradsky isolated *Clostridium pasteurianum*, which is an anaerobic N_2 fixer. Beijerinck isolated *Azotobacter* as a free living aerobic N_2 fixing organism.

Cross inoculation groups of *Rhizobium* (CIG)

It (CIG) refers the groups of leguminous plants that will develop effective nodules when inoculated with the rhizobia obtained from the nodules from any member of that legume group.

I. Rhizobium

Forms nodules on different legume host plants

II. Bradyrhizobium

B. japonicum	Soybean	Soybean
B. spp	Cowpea group	Cowpea group plants

III. Azorhizobium

Stem nodulating – one. Nodulates *Sesbania rostrata.*

IV. Photorhizobia

Nodulantes aeschynomene sp.

V. Sinorhizobium

Sinorhizobium is a fast growing soybean nodulator.

Process of N_2 fixation

The process of N_2 fixation is mediated by the enzyme, called nitrogenase (which mediates the reduction of N_2 to ammonia) first, this enzyme was extracted from the anaerobic di nitrogen fixer *Clostridum pasteurianum*. Latter, this has been isolated from most other N_2 fixing bacteria.

The mechanism of N_2 fixation appears to be quite similar in most N_2 fixing prokaryotes. The enzyme has been fairly well characterized and the enzymes from these different systems share common properties allowing a unified single description of nitrogenase.

Nitrogenase

Nitrogeanse is a functional enzyme which reduces N_2 to ammonia and depends on energy source from ATP. The nitrogenase has two components one containing Mo-Fe, designated as Mo – Fe protein and the other Fe protein. Two components are necessary for the nitrogenase activity.

Mo-Fe protein

Consists of 4 subunits and having the molecular weight of 220,000 or 270,000 daltons and it is the big component.

Fe-protein

Smaller component, contains 2 subunits, molecular weight 60,000 daltons.

Ammonia is the end product of N_2 fixation. The over all reaction is as follows.

$$N_2 + 3H_2 \xrightarrow{\text{ATP}} 2\,NH_3$$

General pathway of N_2 fixation

This process requires a source of ATP and reductants, which are provided by photosynthesis. 16 molecules of ATP are required to fix a molecule of N_2.

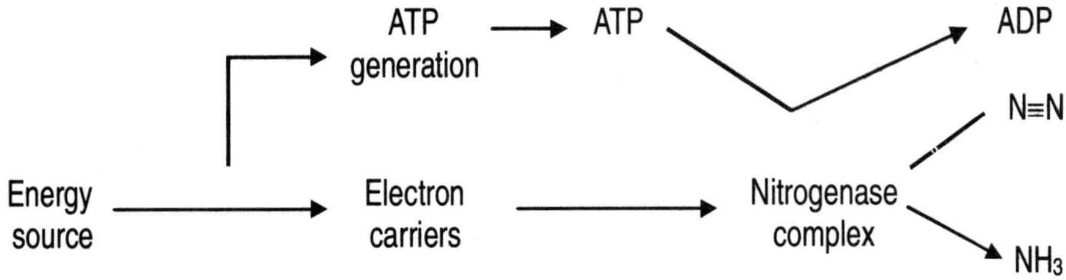

Hydrolysis of ATP into ADP with electron transfer from a reduced electron donor (Ferridoxin, Flavodoxin) is coupled to reduce N_2 to $2NH_3$. The ammonia is the first stable product of fixation and it is assimilated by GS-GOGAT pathway.

Nitrogenase is O_2 labile various protection mechanism are operating in different N_2 fixing systems.

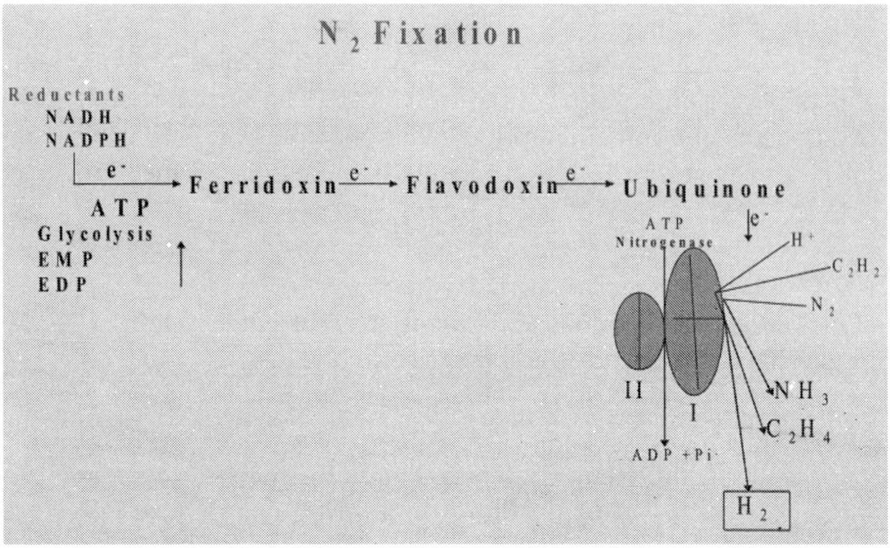

Mechanism

Reduction takes place on the surface of the enzyme

- Six electrons are required to reduce one mole of N to two moles of ammonia.

N_2 + $8H^+$ +$8e^-$ +16 ATP $\rightarrow$ 2 NH_3+16 ADP+ $2H^+$ +16pi

It is postulated that, atoms of N_2 are separated through charge in the valency of metal ion (mo) bound to the enzyme involved in reduction of N_2. For every electron transfer, 4 ATP moles are required.

Hydrogenase : Uptake hydrogenase (HUP^+) converts the hydrogen released during N_2 fixation and cycled back the Hydrogen for energy generation. By this they contribute 9-10% ATP requirement for N_2 fixation process.

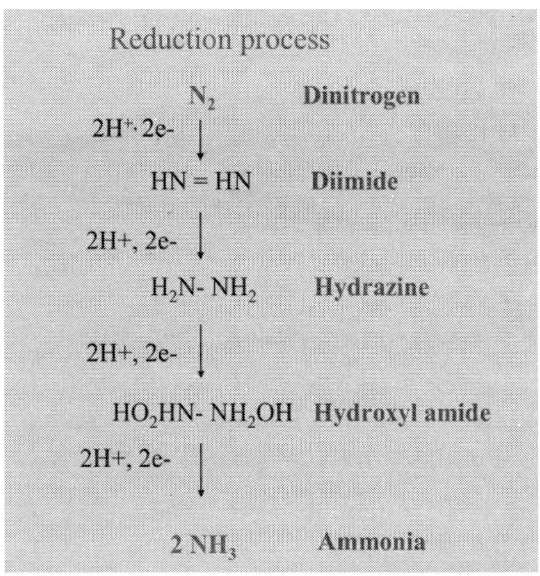

Factors affecting N_2 fixation

1. *Presence of nitrate or ammonium* : More of NO_3/NH_3 reduce N_2 fixation
2. Presence of certain inorganic substances
   Ca, Co, Mo – influence N_2 fixation along with P
3. *Availability of energy source* – addition of C source increases N_2 fixation
4. *pH* : Neutral – favours *Azotobacter* – Acidic- *favours Beijerinkia*
5. *Soil moisture* : Adequate is good for fixation
6. *Temperature* : Mesophilic (30°C)

The energy requirement for BNF is very high and it is a major factor determines the amount of N_2 fixed. In, *Azotobacter* the rate depends on amount of available carbon. In symbiotic N_2 fixers since photosynthesis is the ultimate source of energy the rate of N_2 fixation is influenced by the factors that effect photosynthesis and rate of translocating photosynthates to the N_2 fixing system.

Nitrogenase protection mechansims

1. Respiratory protection, Confirmatory protection

Azotobacter spp O_2 consumption is 1200 U/min. *Azotobacter* has an unusually high rate of respiration. It consumes oxygen so fast that the intracellular concentration of oxygen is kept low enough that nitrogenase is not inhibited. The moment when O_2 enters the cell, it is sent out as CO_2. In order to protect nitrogenase from oxygen, O_2 binding sites are buried within the molecules.

2. Gum protection

Cells are surrounded by copious amounts of exopolysaccharide produced by the cell which protects the nitrogenase from oxygen protection. e.g. *Azotobacter, Beijerinkia, Derxia.*

3. Heterocyst in cyanobacteria

Filamentous cyanobacteria have specialized cells called heterocyst that fix nitrogen. Thick walls of heterocyst protect the diffusion of oxygen into these cells. Spatial separation of N_2 fixing site from photosynthetic sites (vegetative cell) e.g. Anabaena.

4. Leghaemoglobin in Rhizobium *Nodules*

Root nodule tissues and leghaemoglobin protects nitrogenase. An anaerobic environment is maintained in the nodule via the collection of free oxygen by a heme protein, leghaemoglobin (LHb). In addition, oxygen does not diffuse into the nodules easily - a thick layer of cells acts as a barrier to oxygen diffusion.

5. Frankia

Carotenoids and anthocyanin & tannins give protection. Production of vesicles offer protection.

6. Klebsiella

The facultative aerobe *Klebsiella* fixes nitrogen only when it is in an oxygen-free environment.

7. Oxygen insensitive nitrogenase

Discovered in *Streptomyces thermoautotrphicus.*

Overview of Nitrogenase protection from O_2

- Extracellular polysaccharides - Beijerinkia, *Azotobacter*
- Leg haemoglobin – Rhizobium
- Carotenoids, Vesicles – Frankia
- Heterocyst – BGA
- Microaerophillic – *Azospirillum*
- High rate of respiration, confirmatory protection – *Azotobacter*

Losses of N by non biological ways

Leaching

20 to 50%of fertilizer N is lost through leaching. The most striking loss of N occurs in rice soils where more than half of the fertilizer N applied gets lost through leaching.

Volatalization

Another factor is the volatalizaiotn of ammonia in soil which accounts to 5-20%.

Fixation of ammonium

Fixation of ammonium in soils is the minor contributory factor to overall loss of N_2 available for plant growth.

Such losses of N by physical causes and by nitrification and denitirfication process can be controlled by the application of certain chemicals. Some chemicals have been designed to control the rate of release of nutrient from nitrogenous fertilizers, while others retard nitrification in soil by controlling the activity of nitrifying bacteria.

Prevention of N loss

a. Controlled release fertilizers

Urea form Isobutyeldene diurea Crotonilidene diurea S coated urea	Fertilizers, sparingly soluble in water can regulate the release of N from fertilizers

b. Nitrification inhibitors

These are substituted with pyridines, pyrimidines, anilines and isothiocyanates.

Examples:

a. 2 chloro 6 (tricholormethyl) – pyridine – (N serve)

b. 2 amino 4 chloro 6 methyl pyridine – (AM)

N serve inhibits the growth of *Nitrosomonas europea* and *N. agilis.*

The seeds of neem contain lipid associates act as nitrification inhibitors and there by increases the efficiency of urea fertilizers.

Ammonia assimilation

N_2 fixation results in NH_4 formation in heterocysts which reacts with glutamate translocated from vegetative cells to form gutamine with the aid of glutamine synthetase. Most of this glutamine get back into vegetative cells where it is metabolized into glutamate and other amino acids by glutamate synthase or glutamine oxoglutarate amidotransferase (GOGAT) located in vegetative cells. The glutamate formed may eventually form a substrate for more glutamine synthesis or else be transported back into heterocysts. In this way, the first stable product of nitrogen fixation is NH_4^+ and the first organic product of assimilation of NH_4^+ is glutamine followed by glutamate. Alanine may be formed through alanine Dehydrogenase or transamination reaction in some species of BGA and aspartate in other species by transamination reaction probably from glutamate.

Ammonia assimilatory enzymes

- GDH - Glutamate dehydrogenase
- GS - Glutamine synthetase
- GOGAT - Glutamine 2 oxo glutarate amino transfer (Glutamate Synthase)

The sequence of reactions involved

1. Glutamate + NH_4 + ATP $\xrightarrow{GS}$ Glutamine
2. Glutamine + Gluatarate $\xrightarrow{GOGAT}$ Glutamate
3. α - keto glutaric acid $\xrightarrow{GDH}$ Glutamate

Key to enzymes involved in nitrogen assimilation in plant roots

a. Nitrogenase (Legumes only, in association with nitrogen fixing bacteria)
b. Nitrate reductase (NR)
c. Nitrite reductase (NiR)
d. Glutamine synthetase (GS)
e. Glutamate synthase (GOGAT)
f. Glutamate dehydrogenase (GDH)
g. Phosphoenolpyruvate carboxylase (PEPcase)
h. Ferredoxin pyridine nucleotide reductase
i. Asparagine synthetase
j. Glutamate-oxaloacetate aminotransferase

Genetics

Nif genes are responsible for N_2 fixation.

Nif genes are 22, which are located in 7 or 8 clusters (explained in latter pages).

Root nodule bacteria and symbiosis with legumes

One of the most interesting and important plant bacterial interactions is that between leguminous plants and certain gram negative nitrogen fixing bacteria. *Rhizobium, Bradyrhizobium, Sinorhizobium, Mesorhizobium* and *Azorhizobium* are gram negative motile rods. Infection of the roots of a leguminous plant with the appropriate species of one of these genera leads to the formation of root nodules that are able to convert gaseous nitrogen to combined nitrogen, a process called nitrogen fixation. Nitrogen fixation by legume *Rhizobium*, symbiosis is of considerable agricultural importance, as it leads to very significant increases in combined nitrogen in the soil.

As nitrogen deficiencies often occur in unfertilized barren soils, nodulated legumes are at a selective advantage under such conditions. They can grow well in areas where other plants cannot.

Stages in Root nodule Formation

The stages in the infection and development of root nodules are not fairly well understood. They include

1. *Recognition* of the correct partner on the part of both plant and bacterium and *attachment* of the bacterium to the root hairs.
2. *Excretion* of nod factors by the bacterium.
3. *Invasion* of the root hairs by the bacteria forming an infection thread.
4. *Travel* of bacteria to main root via the infection thread.
5. Formation of deformed bacterial cells, *bacteroids*, within the plant cells and development of the nitrogen fixing state.
6. Continued plant and bacterial *division* and formation of the mature root *nodule*.

Nodulation events

Normal root hair

↓

Exudation of organic substances from roots

↓

Accumulation of rhizobia in the rhizosphere

↓

Orientation and binding of rhizobia (A specific adhesion protein called *rhicadhesin* is present on the surface of all species of *Rhizobium* and *Bradyrhizobium*. Rhicadhesin is a calcium-binding protein and may function by binding calcium complexes on the root hair surface. Other substances, such as carbohydrate-containing protein called *lectins*, also play role in plant bacterium attachment)

↓

IAA production by plant roots

↓

Root hair curling and deformation (the root hair curls as a result of the action of substances excreted by the bacterium called nod factor)

↓

Formation of sheperd's crook cells in the legume roots because of the softening of tissues by enzymes

↓

Formation of infection thread by rhizobia

↓

Infection thread containing bacteroids extending into root hair cells

↓

Entry of infection thread into cortex and branching

↓

Root cells adjacent to the root hairs subsequently become infected by rhizobia and nod factors stimulate plant cell division, eventually leading to formation of the nodule

↓

Nodule development

Bacterioids

Bacteriods are specifically referred to a swallon deformed *Rhizobium* cell-found in the root nodule,capable of nitrogen fixation. The Rhizobium bacteria multiply rapidly with in the plant cells and are transformed into swollen, misshapen and branched forms called bacteroids. This is the non motile stage of *Rhizobia*. When the plant dies, the nodules deteriorates, releasing bacteria into the soil. The bacterioid forms are incapable of division, but there are always a small number of dormant rod shaped cells present in the nodule. These now proliferate; using some of the products of the deteriorating nodule as nutrients, and the bacteria can initiate the infection in other roots or maintain a free living existence in the soil.

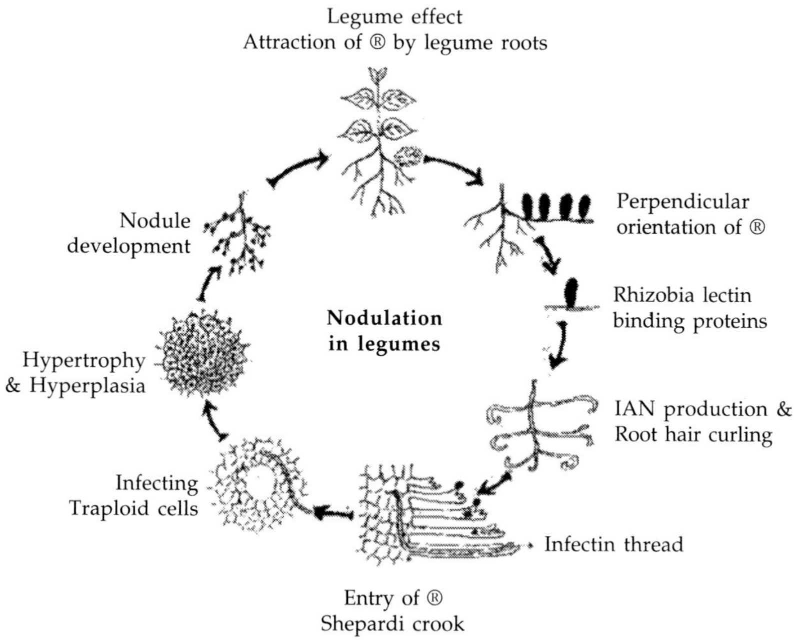

Lectins

Plant proteins which specifically bind to carbohydrate receptors (polysaccharides) in the rhizobial cell are referred as Lectins.

Genetics of nodule Formation

Genes directing specific steps in nodulation of a legume by a strain of *Rhizobium* are called nod genes. Many nod genes from different *Rhizobium* species are highly conserved and are borne on large plasmids called *sym plasmids*. In addition to nod genes which direct specific nodulation events, sym plasmids contain specificity gene, which restrict a strain of *Rhizobium* to a particular host plant. Indeed cross inoculation group specificity can be transferred across species of rhizobia by simply transferring the respective sym plasmid.

In the sym plasmid of *Rhizobium leguminosarum* bio var *viciae, nod* genes are located between two clusters of *nif* genes (genes for nitrogen fixation). Ten *nod* genes have been identified in this species. The *nod* ABC genes are involved in the production of oligosaccharides called *nod factors*, which induce root hair curling and trigger plant cell division, eventually leading to formation of the nodule. In *Rhizobium leguminosarum* bio var *viciae*, the gene *nod* D encodes a regulatory protein; this controls transcription of other *nod* genes.

Nod genes

- Genes directing specific steps in nodulation of a legume by a strain of rhizobium are called Nod genes
- Nod genes are born on large plasmids ,called sym plasmids
- Nod genes are located between two clusters of N_2 fixation genes called Nif genes
- Nod gene consists of 8 genes
- Nod A, B, C, D, E, F, L, M
- Nod D controls the function of all nod genes

Nif Genes

- Genes responsible for N fixation are called *Nif* genes
- 22 genes are involved, arranged in 7 /8 clusters
- Nif Q,B,A,L,F,M,Z,W,V,S,U,X,N,E,Y,T,K,D,H,J
- KDH – control Nitrogenase enzyme complex

Factors affecting nodulation

- Temperature and light
- Combined Nitrogen
- Hydrogen iron concentration
- Mineral nutrition-Co,Mo,P,Ca
- Genetic factors
- Ecological factors
- Salinity and alkalinity

	(his D, G)
Q	Mo uptake or processing
B	FeMoco synthesis or processing
A	Positive regulator
L	Negative regulator
F	Flavodoxin synthesis
M	Processing Fe-protein
Z	?
W	?
V	Processing FeMoco
S	?
U	?
X	?
N	Processing FeMoco
E	Processing FeMoco
Y	?
T	?
K	Synthesis of β-subunit of MoFe-protein
D	Synthesis of α-subunit of MoFe-protein
H	Synthesis of Fe-protein
(C)	?
J	Synthesis of pyruvate-flavodoxin oxidoreductase
	(shi A)

Sequence of nif genes and their probable role in N fixation

❒❒❒

Chapter 18

Microbial Transformation of P and Sulphur

I. Phosphorus cycle

Phosphorus is only second to N_2 as an inorganic nutrient required by both plants and micro organisms. Phosphate constitutes nearly 0.1% of the earth's crust. They occur in soil in inorganic and organic forms.

The inorganic forms are derived from parent rocks or through fertilizer application and manuring with bone meal. They are soluble in water when present as phosphates of Na, K, Ca, Mg *etc*.

The organic phosphorus containing compounds are derived from plants and micro organisms and are composed of nucleic acids, phospholipids, lecithin, phytin and related compounds.

- Phosphorus in phytin, phospholipids and nucleic acids is found as phosphates
- Phytin is the calcium – magnesium salt of phytic acid
- Phospholipids are compounds in which phosphate is combined with a lipid, containing 10% of cell phosphorus.
- Inorganic polyphosphates are quite abundant in certain fungi
- In soil, from15-85% of the total P is organic. Soils rich in organic matter contain abundant organic P.

- Ratios of organic C to P ranges from 100 to 300:1 and N: organic P is 5 to 20: 1

The P Cycle in Soil

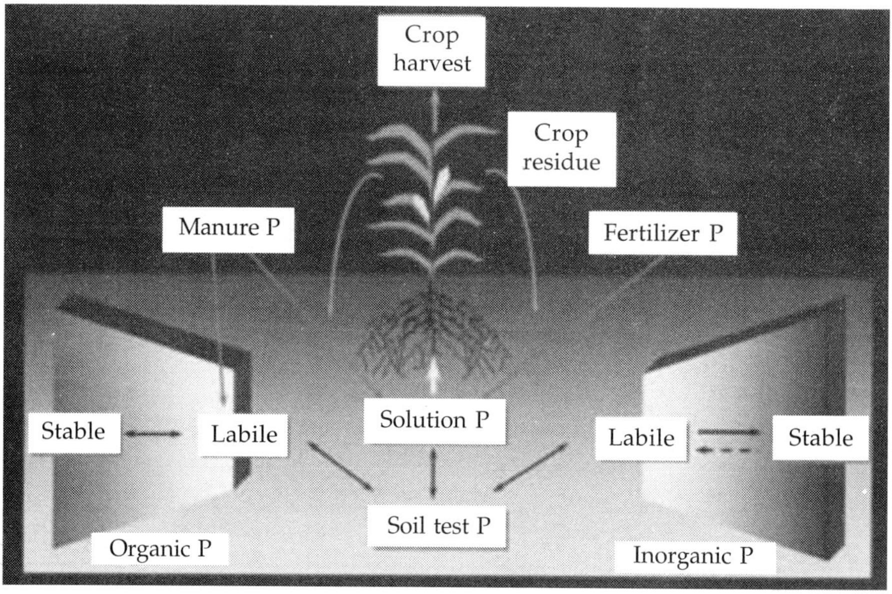

In cultivated soil P present in abundant about 1100 kg/ha but most of them as not available to plants; only about 1% of the total P is in available form.

Role of microbes in P cycle

Microorganisms bring about a number of transformations of the element.

1. Altering the solubility of inorganic compounds of P
2. Mineralization of organic compounds with the release of inorganic phosphate
3. Converting the inorganic, available anion into cell components, an immobilization process (analogous to that occurring with N)
4. Bringing about an oxidation or reduction of inorganic P compounds

Particularly, important to P cycle are the microbial mineralization and immobilization reactions.

1. Solubilization of inorganic phosphorus

Insoluble inorganic compounds of P are largely unavailable to plants, but many microorganisms can bring the PO_4 into solution. P solubilizing microbes are present at 10^5 to 10^7 CFU/ g soil.

Eg: *Pseudomonas striata, Micrococcus, Bacillus sp., Fusarium, Aspergillus* sp, Solubilises calcium salts, iron, aluminum, magnesium & manganese phosphates.

- Micro organism and plant root readily dissolve insoluble PO_4 and make them available to plants.
- P is solubilized by the production of organic acids.
- Solubilization of phosphates by plant roots & micro organism is dependent on soil pH.
- In neutral and alkaline soils having a content of calcium, precipitation of $CaPO_4$ takes place. The acids convert $Ca_3 (PO_4)_2$ to di and monobasic phosphates and releases P to plants.
- The acid soils are generally poor in Ca ions and phosphates and precipitated in the form of ferric or aluminum compounds which are not soluble. There, it is solubilized by the addition of PO_4 solublizing micro organism.
- Phosphorus exists mainly as apatites, with the basic formula $M_{10} (PO_4)_6 X_2$. Commonly the mineral (M) is Ca, less often Al or Fe. The anion (X) is either F^-, Cl^-, OH^- or CO^{2-}_3. Diverse combinations of M and X results in 200 forms of P.

2. Mineralization of organic phosphorus

It is a process of convention of organic forms of phosphorus into inorganic available forms of P. Mineralisation is mediated by phosphatases and phytases. There is a highly significant correlation between the rates of N and P convention to inorganic forms.

Organic form of P is the larger reservoir of P in soil. By the action of bacteria, fungi and actinomycetes, bound element in remains of the vegetation and in soil organic matter is made available to succeeding generations of plants.

Among the organic phosphours compounds, lecithin, nucleic acids and phytin occupy a prominent place. Lecithin contains 9.39 % P_2O_5, 1.6% N and 65.36% C.

- Mineralization is favoured by warm temperature, with the thermophilic range being more favourable than mesophilic range.
- Neutral pH increases PO_4 release, which favours microbial metabolism
- Quantity of substrate ie presence of organic P. If more P, more of mineralization
- Mineralization is mediated by the enzymes called phosphatases. These enzymes cleave phosphorus from more frequently encountered organic substrates.
- Phytases liberates PO_4 from phytic acid or its Ca-Mg, Salt, Phytin. They remove PO_4-s, one at a time, yield penta – tetra, di- and mono PO_4 and then finally free inositol.
- *Bacillus, Pseudomonas, Aspergillus, Penicillium, Rhizopus* can synthesize this enzyme. Mycorrhizal (fungi) are also able to mineralize the organic forms of P and increases P uptake by the plants.

(3) Immobilization

Process of assimilation of P into microbial nucleic acids, phospholipids or other protoplasmic substances is called immobilization. It leads to the accumulation of non utilizable forms of the element.

P accounts for 0.5-1.0% of fungus mycelium and 1.0 to 3.0% of the dry weight of the bacteria and actinomycetes.

(4) Oxidation reduction reactions

Biological oxidation of reduced phosphorus compounds into oxidized state.

Phosphite ($HPO_3^=$) is oxidized to phosphate. A number of hetertrophic bacteria, fungi & actinomycetes utilize phosphite as sole P source. Hypophosphites ($HPO_2^=$) can also be oxidized to phosphate by heterotrophs.

$$HPO_3^= \rightarrow HPO_4^=$$

$$HPO_2^= \rightarrow HPO_4^=$$

Reductive process, reductive pathway has also been functioned. PO_4 is reduced to phosphite and hypophosphite.

$H_3PO_4 \rightarrow H_3PO_3 \rightarrow H_3PO_2$

Clostridium butyricum, E. coli form phosphite and hypophosphite from orthophosphate. It is biochemically analogue to the process of denitirification. Only little information is available about this process.

P exist in an organic form in the protoplasm on the death of living organism, this (P) is changed to inorganic phosphoric acid. This is soon converted into insoluble salts of Ca, Fe, Mg and Al. Phosphorus thus alternates between organic and inorganic, and soluble and insoluble forms. In soluble P is solubilized by various acids produced by micro organism.

Microbial activities involved in the cycling of C, N and P are absolutely essential for maintenance of soil fertility.

II. Sulphur cycle

Sulphur like N, is an essential element for all living systems. Because of its inert nature it is not utilized by plants. To be used, first S has to be oxidized or reduced. In soil, it occurs both in organic (S containing amino acids, vitamins) as well as inorganic forms (Sulphur, sulphates *etc.*).

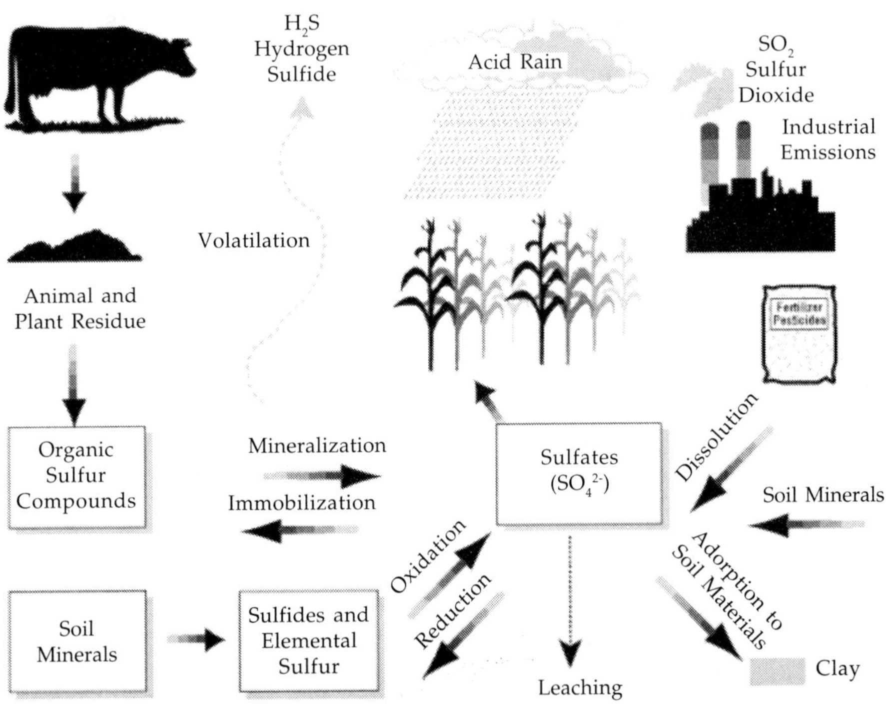

Four distinct transformations are recognized

1. *Decomposition/Mineralization* of larger organic S compounds to smaller units and their conversion into inorganic compounds. The released Sulfates are either absorbed by plants or escape into atmosphere in the form of hydrogen sulphides.
2. *Immobilization* in plants, animals and microbes
3. *Oxidation of inorganic ions and compounds* such as sulphides, thiosulphates, Sulphur into sulfates. Occurs both in aerobic and anaerobic condition.

   Eg. Bacteria -Nonfilamentous forms- *Thiobacillus*, Filamentous forms – *Beggiatoa,Thiothrix* and *Thioploca*

   Fungi and actinomycets - *Aspergillus, Penicillium* and *Microsporium*

   Importance of Thiobacillus

   - Produces Sulphuric acid ,lower down the soil pH – Hence used in controlling plant disease

     Apple and Potato scab – caused by *Streptomyces scabis* ,Sweet potato rot – caused by *S. ipomea*

     S+ *Thiobacillus* application is used for the control
   - Used in remediation of alkali soil
   - Increases the solubilization of other nutrients (P,K,Ca,Mn,Al and Mg)
   - Used in preparation of biosuper- Rock phosphate + *T.thiooxidans* and S— (First prepared in Australia)
   - Used in Lipman's process of compost preparation

     Soil + manure + elemental S + rock phosphate combinations produces enriched compost
4. *Reduction* of Sulphates into S or H_2S under anaerobic condition. Eg. *Desulfovibrio desufuricans*. Reduces the availability of S for plant nutrition.

❒❒❒

Chapter 19

Biofertilizers and their Importance

Biofertilizers

Biofertilizers are defined as preparations containing living cells or latent cells of efficient strains of microorganisms that help crop plants uptake of nutrients by their interactions in the rhizosphere when applied through seed or soil. They accelerate certain microbial processes in the soil which augment the extent of availability of nutrients in a form easily assimilated by plants.

Use of biofertilizers is one of the important components of integrated nutrient management, as they are cost effective and renewable source of plant nutrients to supplement the chemical fertilizers for sustainable agriculture. Several microorganisms and their association with crop plants are being exploited in the production of biofertilizers. They can be grouped in different ways based on their nature and function.

I. N_2 fixers

a. Free living : Aerobic – *Azotobacter, Beijerinckia, Anabaena*
Anaerobic – *Clostridium*
Faultative anaerobic – *Klebsiella*

b. Symbiotic : *Rhizobium, Frankia, Anabaena azollae*

c. Associative symbiotic : *Azospirillum*

d. Endophytic : *Gluconacetobacter, Burkholdria*

II. Phosphorus solubilizers

Bacteria - *Bacillus megaterium* var. *phosphaticum*
B. subtilis, B. circulans
Pseudomonas striata

Fungi : *Penicillium* sp.
Aspergillus awamori

III. P mobilizers

Arbuscular mycorrhizal fungi, Ectomycorrhizal fungi, Ericoid mycorrhiza, Orchid mycorrhiza

IV. Silicate and Zinc solubilizers

Bacillus mucilaginous

V. Plant growth promoting Rhizobacteria

Pseudomans spp., *Methylobacterium* and many more.

VI. K solubilizers

Frateuria aurentia

Importance of Biofertilizers

Biofertilizers are known to make a number of positive contributions in agriculture.

- Supplement fertilizer supplies for meeting the nutrient needs of crops.
- Add 20-200 kg N/ha (by fixation) under optimum conditions and solubilise/mobilise 30-50 kg P_2O_5/ha.
- They liberate growth promoting substances and vitamins and help to maintain soil fertility.
- They suppress the incidence of pathogens and control diseases.
- Increase the crop yield by 10-50%. N_2 fixers reduce depletion of soil nutrients and provide sustainability to the farming system.

- Cheaper, pollution free and based on renewable energy sources.
- They improve soil physical properties, tilth and soil health.

A. Nitrogen Fixers

1. Rhizobium

Rhizobia are soil bacteria, live freely in soil and in the root region of both leguminous and non-leguminous plants. However they enter into symbiosis only with leguminous plants, by infecting their roots and forming nodules in them. Non legume nodulated by Rhizobia is *Trema* or *Parasponia* sp.

The nodulated legumes contribute a good deal to the amount of N_2 fixed in the biosphere, (50-200 kg N/ha) varied with crops.

Clover - 130 kg N/ha

Cowpea - 62 – 128 kg N/ha

Beijerinck first isolated and cultivated a microorganism from the roots of legumes in 1888 and he named this as *Bacillus radicola* and later modified as *Rhizobium.*

Legume plants fix and utilize this N by working symbiotically with *Rhizobium* in nodules on their roots. The host plants provide a home for bacteria and energy to fix atmospheric N_2 and in turn the plant receives fixed N_2 (as protein).

Cross inoculation group (CGI)

Genera / species	Principal and other reported hosts
Rhizobium	
R.etli	*Phaseolus vulgaris, Mimosa affinis*
R.galegae	*Galega orientalis, G.officinalis*
R.gallicum	*Phaseolus vulgaris, Leucaena, Macroptilium, Onobrychis*
R.giardini	*Phaseolus vulgaris, Leucaena, Macroptilium*
R.hainanense	*Desmodium sinuatum, Stylosanthes, Vigna, Arachis, Centrosema*
R.huautlense	*Sesbania herbacea*
R.indigoferae	*Indigofera*
R.leguminosarum	
bv trifolii	*Trifolium*
bv viciae	*Lathyrus, Lens, Pisum, and Vicia,*
bv.phaseoli	*Phaseolus vulgaris*

R.mongolense	*Medicago ruthenica, Phaseolus vulgaris*
R.sullae	*Hedysarum coronarium*
R.tropici	*Phaseolus vulgaris, Dalea, Leucaena, Macroptilium, Onobrychis*
Mesorhizobium	
M.amorphae	*Amorpha fruticosa*
M.chacoense	*Prosopis alba*
M.ciceri	*Cicer arietinum*
M.huakuii	*Astragalus sinicus, Acacia*
M.loti	*Lotus corniculatus*
M. mediterraneum	*Cicer arietinum*
M.plurifarium	*Acacia senegal, Prosopis juriflora, Leucaena*
M.tianshanense	*Glycyrrhiza pallidflora, Swansonia, Glycine, Caragana, Sophora*
Sinorhizobium	
S.abri	*Abrus precatorius*
S.americanus	*Acacia spp.*
S.arboris	*Acacia senegal, Prosopis chilensis*
S.fredi	*Glycine max*
S.indiaense	*Sesbania rostrata*
S.kostiense	*Acacia senegal, Prosopis chilensis*
S.kummerowiae	*Kummerowia stipulacea*
S.meliloti	*Medicago, Melilotus, Trigonella*
S.medicae	*Medicago truncatula, M. polymorpha, M.orbicularis*
S.morelense	*Leucaena leucocephala*
S.sahelense	*Acacia, Sesbania*
S.terangae	*Acacia, Sesbania*
Azorhizobium	
A.caulinodans	*Sesbania rostrata*
Allorhizobium	
A.undicola	*Neptunia natans, Acacia, Faidherbia, Lotus*
Bradyrhizobium	
B.elkanii	*Glycine max*
B.japonicum	*Glycine max*
B.liaoningense	*Glycine max*
B.yuanmingense	*Lespedeza, Medicago, Melilotus*

Description and Characteristics of *Rhizobium*

Classification

1. Family	:	Rhizobiaceae
2. Genus	:	*Azorhizobium*-for stem nodulation *Bradyrhizobium* *Rhizobium* *Sinorhizobium*

Morphology

1. Unicellular, cell size less than 2μ wide, short to medium rod, pleomorphic.
2. Motile with Peritrichous flagella
3. Gram negative
4. Accumulate PHB granules.

Physiology

1. Nature	:	Chemoheterotrophic, symbiotic with legume
2. C source	:	Supplied by legume through photosynthates, monosaccharides, disaccharide
3. N source	:	Fixed atmospheric N_2
4. Respiration	:	Aerobic
5. Growth	:	Fast (*Rhizobium*), slow (Bradyrhizobium)
6. Doubling time	:	Fast growers – 2-4 hrs Slow growers – 6-12 hrs
7. Growth media	:	Yeast Extract Mannitol Agar Medium (YEMA)

Contribution

1. Direct contribution of N symbiotically with legumes.
2. Residual nitrogen benefit for the succeeding crop.
3. Yield increase is by 10-35%.
4. Improve soil structure.
5. Produces exopolysaccharides.
6. Produces plant growth hormone.

Recommended for legumes (Pulses, oilseeds, fodders)

Promising strains : NGR 6, NC 92, CC 1, CRR 6, CRU 14, COBE 13.

2. *Azotobacter*

It is a free living N_2 fixer, the cells are not present on the rhizoplane, but are abundant in the rhizosphere region. It is classified under the family Azotobacteriaceae. It requires more of organic matter and depend on the energy derived from the degradation of plant residues. Beijerinck was the first to isolate and describe *Azotobacter*.

Species

Cell size, flagellation, pigmentation and production of extracellular slime are considered as diagnostic features of these bacteria in distinguishing species.

A. chroococcum	:	Black to brown insoluble pigment.
A. vinelandii, A. paspali, A. agilis	:	Green fluorescent and soluble pigments
A. beijerinckii	:	Yellow to light brown insoluble pigments
A. macrocytogenes	:	Pink soluble pigments
A. insignis	:	Yellow brown pigments

Azotobacter cells are polymorphic, gram negative, form cyst and accumulate Poly Beta hydroxy butyric acid and produces abundant gum.

Morphology

Cell size	:	Large ovoid cells, size 2.0-7.0 x 1.0-2.5 μ
Cell character	:	Polymorphic
Gram character	:	Negative

Physiology

1. Nature	:	Chemoheterotrophic, free living
2. C source	:	Mono, di saccharides, organic acids

3. N source : N_2 through fixation, amino acids, NH_4^+, NO_3^-
4. Respiration : Aerobic
5. Growth : Ashby / Jensen's medium
6. Doubling time : 3 hours

Benefits

- Ability to fix atmospheric N_2 – 20-40 mg BNF/g of C source in laboratory equivalent to 20-40 kg N/ha.
- Production of growth promoting substances like vitamin B, Indole acetic acid, GA.
- Ability to produce thiamine, riboflavin, pyridoxin, cyanogobalanine, nicotinic acid, pantothenic acid, etc.
- Biological control of plant diseases by suppressing *Aspergillus, Fusarium*.

Recommended for Rice, wheat, millets, cereals, cotton, vegetables, sunflower, mustard and flowers.

3. Azospirillum

Azospirillum was isolated by *Beijerinck* (1922) in Brazil from the roots of *Paspalum* and named it as *Azotobacter paspali* and later named as *Spirillum lipoferum*. Dobereiner and Day (1976) reported the nitrogen fixing potential of some forage grasses due to the activity of *A. lipoferum* in their roots. Dobereiner coined the term "*Associative symbiosis*" to denote the occurrence of N_2 fixing *spirillum* in plants. Taxonomy was re-examined and *Tarrand* and co-workers in 1978 designated this organism as *Azospirillum*.

It is an aerobic or micro aerophilic, motile, gram negative bacterium. Non spore former and spiral shaped bacterium, inhabiting the plant roots both externally and internally. Being a micro aerophilic organism, it can be isolated on a semi solid malate medium by enrichment procedures.

Classification

Family – *Spirillaceae*

Species : (7)

1. *A. brasilense*
2. *A. lipoferum*

3. *A. amazonense*
4. *A. halopreferens*
5. *A. irakense*
6. *A. dobereinerae*
7. *A. largimobilis*

Morphology

1. Cell size : Curved rods, 1 mm dia, size and shape vary
2. Accumulate : PHB
3. Gram reaction : Negative
4. Development of white pellicles : 2-4 mm below the surface of NFB medium

Physiology

1. Nature : Chemoheterotrophic, associative
2. Sole carbon source : Organic acids, L-arabinose, D-gluconate, D-fructose, D-glucose, sucrose, Pectin
3. N source : N_2 through fixation, amino acids, N_2, NH_4^+, NO_3^-
4. Respiration : Aerobic, Microaerophilic
5. Growth media : NFBTB (NFB) medium Nitrogen free Bromo thymol bline medium
6. Doubling time : 1 hr in ammonia containing medium 5.5-7.0 hrs in malate containing semisolid medium

Contributions

- Contribution through BNF
- Production of Plant Growth Promoting substances by bacteria
- Increases root hair development, biomass.
- Enhance the production of PGP substances by plant results in

  Morphological changes in root cells.

  Increased activity of IAA oxidase

  Increase in endogenous IAA

- Increased mineral and water uptake, root development, vegetative growth and crop yield.
- Out competes in the rhizosphere with other harmful microorganism.
- Production of polyamines and amino acids.
- Increased extrusion of protons and organic acids in plants.

Benefits

1. Promotes plant growth.
2. Increased mineral and water uptake, root development, vegetative growth and crop yield.
3. Inoculation reduced the use of chemical fertilizers (20-50%, 20-40 kg N/ha)
4. Increases cost benefit ratio.
5. Reduces pathogen damage.
6. Inhibit germination of parasitic weeds.
7. Restoration of arid zone, margine mangrove ecosystem.
8. Reduces humic acid toxicity in compost.

Recommended for rice, millets, maize, wheat, sorghum, sugarcane and co-inoculant for legumes.

4. Gluconacetobacter diazotrophicus

It is an endophytic N_2 fixer and form to occur on large numbers in roots, stem and leaf of sugarcane and other sugar rich crops. It was first isolated from sugarcane. *Cavalcante and Dobereiner* in 1988, reported this new endophytic N_2 fixer and recently renamed as *G. diazotrophicus*. It can tolerate upto 30% sucrose concentration and pH upto 3.0. Optimum sucrose concentration is 10-15%. Does not grow at pH 7.0. Optimum is 5.5. Produce surface yellow pellicle on semisolid medium.

Benefits

- Fixes atmospheric N_2
- Production of PG hormones (GA, IAA is double than *Azospirillum*).

- Suitable for sugar rich crops with acidic pH
- Carry out P & Zn solubilisation to some extent

5. Azorhizobium

This genera can produce stem nodules. Species is *Azorhizobium caulinodans.* Stem nodulation has been reported in 3 genera of legumes: *Aeschynomene, Neptunia* and *Sesbania.*

Stem nodulating *Rhizobium* comprises both fast and slow growing types having the generation time of 3-4 hr and 10 hrs respectively. Those nodulate *Aeschynone* can cross inoculate with *S. rostrata.*

- fix N_2 in free living conditions without differentiating into bacteroids.
- have O_2 protection mechanisms, to fix N_2 under free living conditions.
- Mode of entry is through lateral root cracks. No infection thread is formed during colonization.
- Form both stem and root nodules in *S. rostrata.*
- Gram negative, motile rods.
- Produces root nodules in rice, wheat.

6. Algal Biofertilizers

The agronomic potential of cyanobacterial N_2 fixation in rice fields was recognised in India during 1939 by **De** who attributed the natural fertility of tropical rice fields to N_2 fixing blue green algae. The rice field ecosystem provides an environment favourable for the growth of blue green algae with respect to their requirements for light, water, high temperature and nutrient availability.

Blue Green Algae / Cyanobacteria

Blue green algae are a group of prokaryotic, photo synthetic microscopic plants, named as Cyanobacteria. They have striking morphological and physiological similarities with bacteria and hence also called as cyanobacteria. Some of them are able to fix N_2. They can be divided into two major groups based on growth habitats.

a) Heterocystous – aerobic forms
b) Aerobic unicellular forms
c) Non-heterocystous, filamentous, micro aerophilic forms.

N_2 fixing species of both unicellular and filamentous groups are found in paddy fields, but the predominant ones are the *heterocystous filamentous forms*.

Cyanobacteria are not restricted to permanently wet habitats, as they are resistant to desiccation and hot temperatures, and can be abundant in upland soils. However wet paddy soils and overlying flood waters provide an ideal environment for them to grow and fix N_2.

Natural distribution

BGA are cosmopolitan in distribution. They are widely distributed in tropical zone. Free living cyanobacteria can grow epiphytically on aquatic and emergent plant as well as in flood water or on the soil surface. Heterocystous cyanobacteria formed less than 10% of the population of eukaryotic green algae. The abundance increased with the amount of available phosphorus and with the pH value over the range 4-6.5. In rice soil, population ranges from 10^6-10^7 cfu g^{-1} soil.

The main taxa of N_2 fixing cyanobacteria

Group	Genera	DNA (mol % GC)
Group-I. Unicelluar: single cells or cell aggregates	*Gloeothece, Gloeobacter,Synechococcus, Cyanothece, Gloeocapsa, Synechocystis, Chamaesiphon, Merismopedia*	35-71
Group-II. Pleurocapsalean: reproduce by formation of small spherical cells called baeocytes produced through multiple fission.	*Dermocarpa, Xenococcus, Dermocarpella, Pleurocapsa, Myxosarcina, Chroococcidiopsis*	40-46
Group-III. Oscillatorian: filamentous cells that divide by binary fission in a single plane.	*Oscillatoria, Spirulina, Arthrospira, Lyngbya, Microcoleus, Pseudanabaena.*	40-67
Group-IV. Nostocalean: filamentous cells that produce heterocysts	*Anabaena, Nostoc, Calothrix, Nodularia, Cylinodrosperum, Scytonema*	38-46
Group-V.Branching: cells divide to form branches	*Fischerella, Stigonema, Chlorogloeopsis, Hapalosiphon*	42-46

The N_2 fixing forms generally have a specialized structure known as *heterocyst*. The BGA have minimum growth requirement need only diffused light, simple inorganic nutrients and moisture. The heterocysts which are modified vegetative cells, because of their thick walls and absence of photosystem II in photosynthesis, act as ideal sites for N_2 fixation under aerobic conditions. Although the nitrogenase is present in vegetative cells, it remains inactive because of the presence of oxygenic photosynthesis. They build up natural fertility (C, N) in soil.

Contributions of algal biofertilizer

- Important component in organic farming.
- Contribute 20-25 kg N ha^{-1}.
- Add organic matter to the soil.
- Excrete growth promoting substances.
- Solubilize insoluble phosphates.
- Improve fertilizer use efficiency of crop plants.
- Improve physical and chemical properties of soil.
- Reduce C:N ratio.
- Increase the rice yield by 25-30%.
- Cyanobacteria are more compatible with nitrate N than ammonium N.

It increases fertilizer use efficiency of the crop plants through exudation of growth promoting substances and preventing a part of applied fertilizer N from being lost.

Composite BGA inoculum : The blue green algal culture's composite inoculum consisting of *Nostoc, Anabaena, Calothrix, Tolypothrix, Plectonema, Aphanotleca, Gleocapsa, Oscillatoria, Cylindrospermum, Aulosira and Scytonema.*

B. Phosphate solubilisers

Though most soils contain appreciable amounts of inorganic P, most of it being insoluble forms, cannot be utilized by crops unless they are solubilzied. Phosphate solubilizing microorganisms not only able to solubilize insoluble forms of inorganic P but are also capable to mineralize organic forms of P, thus improving the availability of native soil P.

Mechanism of P solubilization

Different mechanisms were suggested for the solubilization of inorganic phosphates.

- Production of organic acids
- Chelating effect (Siderophores, bind iron tightly to prohibit its reaction with soluble phosphate and rather help release PO_4 fixed as ferric phosphate. It is important in acid soils, where ferric PO_4 is one of the major forms)
- Production of inorganic acids
- Hydrogen sulphide production (H_2S)
- Effect of carbon dioxide
- Proton extrution

Microbes involved

Bacillus megaterium var. *phosphaticum, B. subtilis, B. circulans, Penicillium striata, Aspergillus awamori, Aspergillus niger, Streptomyces*

C. PO_4 mobilizer – Mycorrhiza

Mycorrhiza (fungus root) is the mutualistic association between plant roots and fungal mycelia. *Frank* in 1885 gave the name *"mycorrhiza"* to the peculiar association between tree roots and ectomycorrhizal fungi. This association is characterized by the movement of plant produced carbon to fungus and fungal acquired nutrients to plants.

Types of mycorrhiza

Based on the association with crop plants, it has been grouped into the following types.

1. Ectomycorrhiza

- The fungal hyphae form a *sheath or mantle* on outside the root which increases the surface area of absorbing roots and offers protection to the roots.
- Within the root form an extensive network called the *Hartignet* in the intercellular spaces of the epidermis and cortex. Hartignet can act as storage and transport organ for P.

- *No intracellular* penetration into epidermal or cortical cells occurs
- Ectomycorrhizae are common on trees, including members of the families pinaceae (Pin, Fir, Spruce, Larch, Semlock), Fagaceae (Willow, Poplar, Chesnut), Betulaceae (Birch, Alder), Salicaceae (Willow, Poplar) and Myrtaceae.
- The fungi forming Ectomycorrhizal association are coming under Basidiomycotina and Ascomycotina. eg: *Laccaria laccata, Suillus, Rhizopogan, Amanita*

2. Endomycorrhizae

- This fungus do not form an external mantle but lives within the root.
- They produce an internal network of hyphae between cortical cells that extends out into the soil, where the hyphae absorb mineral salts and water.
- In all forms, hyphae runs between and inside the root cells
- Endomycorrhizae consist of three sub groups

Arbuscular mycorrhiza	-	Associated with most of the plant families
Ericoid mycorrhiza	-	Associated with some species of Ericaceous plants
Orchid mycorrhiza	-	Associated with orchid plants

a. Arbuscular Mycorrhizal fungi

AM, an endomorphic mycorrhizae formed by the aseptate phycomycetous fungi and are associated with majority of agricultural crops, growing under broad ecological range. 150 species of AMF are known.

Class	:	Zygomycotina
Order	:	Endogonales
Family	:	Endogonaceae

Colonization Process

Colonization occurs under two phases: (1) Extra matrical phase and (2) Intra radical phase.

Extra matrical phase: Events occurring outside the root after the germination of chlamydospores. Mycelium explores larger soil volume. Fungal growth can be 80-130 times the length of root. Extra matrical hyphae (EMH) are larger in diameter than inner hyphae. Once the fungus recognises the plant, appresorium is formed in the host roots and penetration occurs via the appresorium. EMH ends with resting spores in soil.

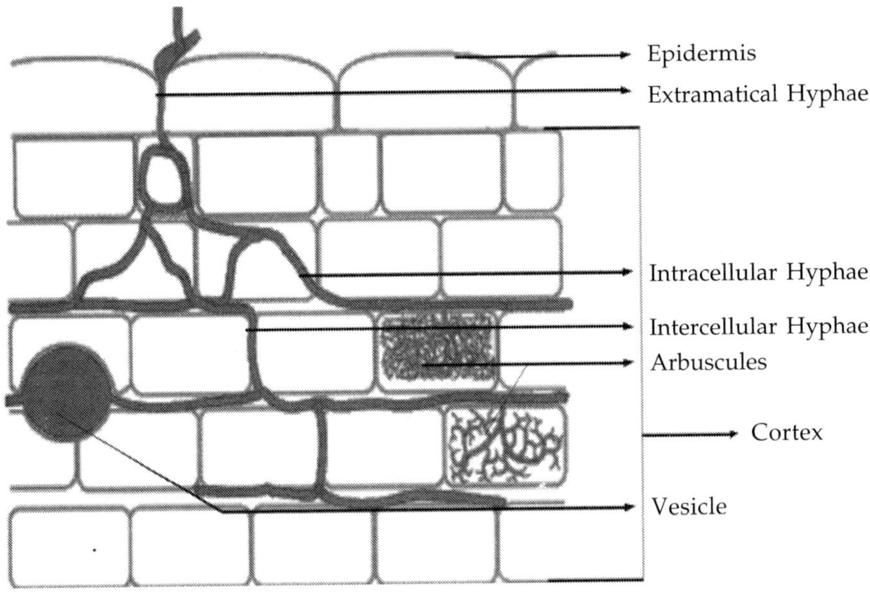

Intra radical phase: Events occurring inside the root cortex. After penetrating the cortex, the fungus may produce intercellular as well as intracellular hyphae in the cortical cells. Forms two morphological structures namely arbuscules and vesicles inside the cortical cells.

Arbuscules: are the first formed structures after the hyphal entry into the cortical cells. Arbuscules are the fine dichotomously branched hyphal filaments look like little trees. Arbuscules start to form approximately 2 days after penetration. They are considered as the major site of *exchange between the fungus and host root*. They are short lived (*4-13 days*) and degenerate.

Vesicles: Following the formation of arbuscules, some species of fungi also form vesicles in the roots. Terminal or intercallery hyphal swellings of the hyphae called vesicles. Vesicles contain lipids and cytoplasm. They act as *P storage* organ and they *ever be present* in the root. Size of the vesicles is about 30-100 µm. In vesicles P can be accumulated as *polyphosphates*.

Contributions

The beneficial effect on plant growth and yields following inoculation with AM is attributed to :

(i) Improved mineral nutrition, especially P (P, Zn, Cu, K, S, NH_4)
(ii) Mobilization of nutrients through greater soil exploration
(iii) Protection of host roots against pathogen infection
(iv) Improved water relation
(v) Better tolerance to stress like salinity, heavy metal
(vi) Protection against transplantation shock.

Reasons for Enhanced P uptake by AM Fungi

- Physical exploration of soil.
- Higher affinity towards P
- Lower threshold concentration for hyphal absorption of P
- Rhizosphere modification called Mycorrhizosphere
- Differences in anion and cation absorption due to exudation pattern.
- Siderophore production.
- Selective stimulation of microorganisms in the rhizosphere.
- Increased hyphal area for absorption (EMH).
- Absorb and transport P beyond the depletion zone around the root.
- P absorption by EMH is 1000 times faster than normal hyphae and 3-4 times greater.

Disease resistance

- Resist the parasitic invasion and minimises the loss.
- Mycorrhizal roots harbour more actinomycetes which contributes for antibiotic substances
- Mycorrhizal roots have elevated levels of phenols, while offers resistance to fungal hydrolytic enzymes.
- Mycorrhizal infection stimulates biosynthesis of phytoalexins.

❑❑❑

Part 3

Industrial Microbiology

Chapter 20

Definition and Scope of Industrial Microbiology

Industrial microbiology deals with all forms of microbiology which have an economic aspect. Industrial microbiology concerns itself with (i)the isolation and description of microorganisms from natural environments, such as soil or water, (ii) with the cultural conditions required for obtaining rapid and massive growth of these organisms

in the laboratory (iii) performing these fermentations in large scale cultural vessels commonly known as fermenter. The detection and assay products of microbial activity, the recovery, chemical purification, packaging, and marketing of fermentation products are the various activities of industrial microbiology.

Industrial microbiology is the study that deals with the ability of microorganisms to convert inexpensive raw materials, or subtrates, to economically valuable organic cornpounds obviously is of considerable concern to the industrial microbiologist. Also of great interest is the economic value of the microbial cells themselves, and of the intracellular and extracellular enzymes/metabolites elaborated by the organisms.

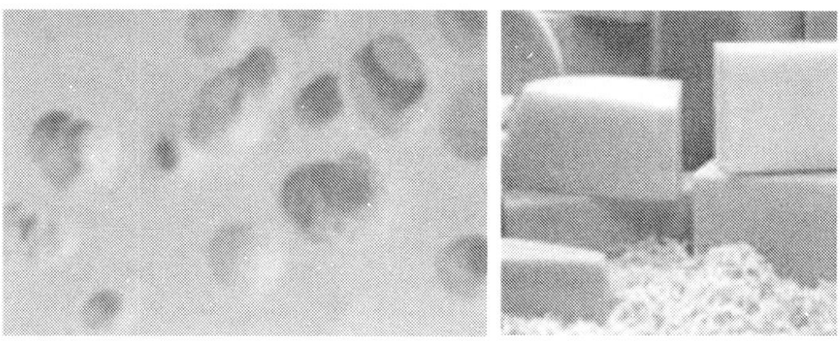

Cheese

Products of microbial activity having potential and commercial importance

1. Substrates- to economically valuable organic compounds (bioconversion products)

- Beverages: wine, beer, and distilled spirits
- Foods: cheeses, fermented milks, pickles, sauerkraut, soy sauce, yeast, bread, vinegar, mushrooms, and acidulants such as critric acid

2. Intracellular and extracellular compounds elaborated by the organisms (metabolites)

- Amino acids: L-glutamic acid and L-lysine
- Steroids
- Vitamins and other growth stimulants: B12, riboflavin, vitamin A and the gibberellins
- Enzymes: amylases, proteases, pectinases, invertase, etc.

- *Antibiotics*: streptomycin, penicillin, the tetracyclines, erythromycin, polymyxin, bacitracin, etc.
- *Organic solvents*: acetone, butanol, ethanol, amyl alcohol, etc
- *Organic acids*: lactic, acetic, citric, gluconic, buytyric, itaconic, kojic, etc
- *Flavoring agents*: monosodium glutamate and nucleotide

3. Microbial cells themselves (cell biomass)

- Baker's yeast
- Food and feed yeast
- Biofertilizers
- Bacterial insecticides/
- Biocontrol agents

Apart from this, following are the other notable industrially important fermentations :

Ensiling, composting, anaerobic digestion, soil and groundwater remediation, bioleaching, microbial enhanced oil recovery, microbial metals recovery, waste treatment, coal and oil desulfurization.

Fermentation

Any process mediated by or involving microorganisms in which a product of economic value is obtained, is called *fermentation*. The word fermentation is derived from the Latin verb *fervere*, 'to boil' thus describing the appearance of the yeast or extracts of fruit or malted grains. In fact, from an industrial microbiology standpoint fermentation means almost any process mediated by or involving microorganisms in which a product of economic value accrues.

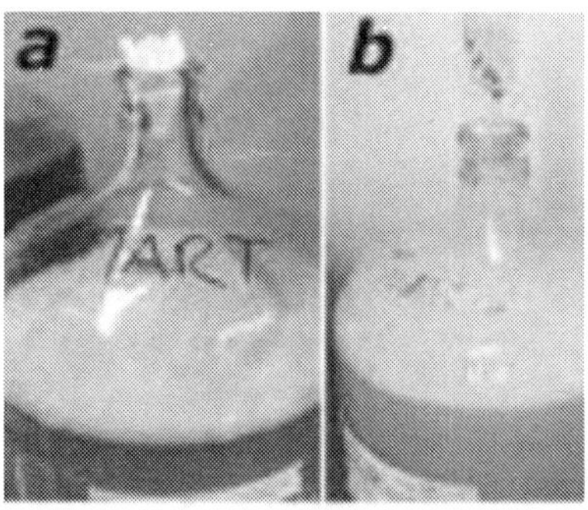

The Fermentation

The basic requirements for industrial fermentation include an organism, the medium and a process for product recovery.

1. *The organism* : A suitable organism is sought by screening from a population or creating specific strains of microorganisms that will yield high quantities of the desired product by genetic engineering. The strain thus selected should have rapid growth, stable characters, non pathogenic and non producer of any unwanted byproduct or toxin.
2. *The medium* : It should contain least expensive compounds, readily available and high yielding components. Mostly those nutrient containing wastes of certain industries are used like whey from dairy industry and waste sulphite liquor from paper industries.
3. *The product and Recovery* : fermentors with huge capacities should be designed in such a way that it should provide all environmental conditions for high yield of product and easy recovery of the product.

□□□

Chapter 21

Historical Development of Industrial Microbiology

Industrial microbiology is a relatively new science, as is all of microbiology. As a science, it is a little more than 100 years old; as an art, it dates back into antiquity. However, a few of the present-day practices of industrial microbiology are more of an art than a science. The basic concepts of industrial microbiology and the academic undertones associated with it will be better understood if we first consider some of its historical aspects.

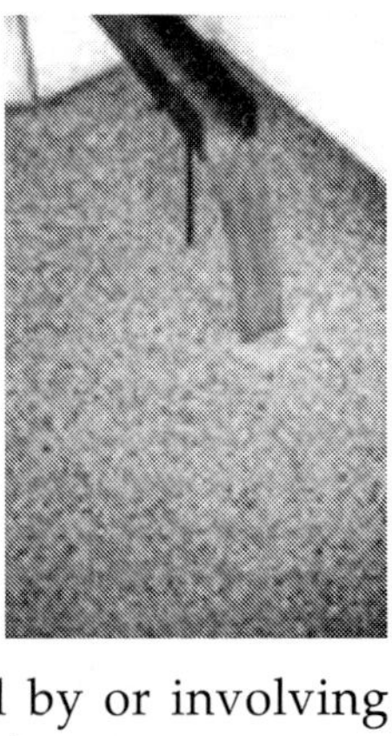

Originally, fermentation referred to the bubbling observed when sugar and starchy materials underwent a transformation to yield alcoholic beverages. Later, this term was applied to the process in which alcohol was formed from sugar, regardless of whether the causative agent was or was not biological. He considered fermentation to apply to those anaerobic reactions through which microorganisms obtained energy for growth in the absence of oxygen. In fact, from an industrial microbiology standpoint fermentation means almost any process mediated by or involving microorganisms in which a product of economic value accrues.

Man practiced these processes back into antiquity, and the first usage of most of the processes is, in large part, unknown. Honey probably was available even to the earliest civilizations without requiring specific effort for its production and, therefore, fermented honey, or mead, may have provided one of the earliest alcoholic beverage fermentations for man. Manna underwent fermentation in a manner analogous to that of honey to yield a fermented drink. Other early fermentation processes practiced by man included the leavening of bread, the retting of flax, the production of various alcoholic beverages (including beer, wine, and mead), the production of vinegar by allowing wine to sour and the production of various fermented foods and milks.

Contribution of **Anton Von Leeuwenhoek** (1680) is of particular importance in relation to industrial microbiology. He described beer yeast as being very small spherical or oval granules by observing under very primitive microscope which he had developed by grinding lenses. However, his observations did not experimentally establish that these minute forms were living entities.

Francesco Redi's experiments in the 17th century demonstrated that the macro concepts of spontaneous generation were fallacious. He proved this by conducting an experiment by placing a piece of meat in a flask and covered it with a mesh and observed that there was no spontaneous maggot formation from the meat. Though he could not explain that normally maggots arise from the egg mass that are laid by some flies when the meat is exposed, he could put one step ahead in breaking the theory of spontaneous generation. Anyway he was not able to explain this at microlevel.

Lazaro Spallanzani attempted to disprove spontaneous generation by hermetically sealing infusions of seeds in flasks, followed by heating of the flasks for varying periods of time in a boiling water bath. His experiments demonstrated the difference in heat resistance of various groups of organisms and prolonged period of heating is necessary for sterilization. Exposure of the medium to air caused microbial growth which made the supporters of spontaneous generation to conclude that heat killed the microbes in medium but reentry of atmospheric air into the flask cause the spontaneous formation of life from the medium components.

Theodore Schwann (1837) attempted to answer the above criticism of Spallazani's experiments. He demonstrated that a sterilized meat broth did not undergo putrifaction if fresh heated air was introduced into the medium. He considered that alcoholic fermentations might be caused by living organisms that destroyed the sugar in the process producing alcohol and CO_2. Microscopes by this time had developed to the point that Schwann was able to observe the yeast and, in fact, was the first to observe and describe yeast in the process of growing. He observed budding, and concluded that the organism was fungal-like. He also stated that fermentation is the decomposition wherein the sugar fungus (yeast) utilizes sugar for its growth and the substances that do not go into the plant (yeast) are converted to alcohol.

Charles Cagniard-Latour (1838) confirmed and extended Schwann's descriptions of the growth and morphology of yeast cells. **Gay-Lussac** had previously demonstrated that oxygen was required for initiation of the alcoholic fermentation but that it was not required for the further progress of the fermentation.

Justis Liebig (1839), used his influence and scholarly status to proclaim that all of the reported apparently living activity observed by these and other workers was really the result of chemical and physical reactions going on in the medium. Beer yeast and all other organic materials undergoing decomposition really were merely unstable chemical substances.

Then **Louis Pasteur** entered into the picture. Pasteur's paper published in 1857 represented the beginning of the science of microbiology. This was Pasteur's first reported study of fermentation, and it was concerned not with the alcohol fermentation but with the lactic acid fermentation. He described a medium for lactic acid production. he separated the bacteria from the fermentation medium by washing the cells, but was not able to get the cells into pure culture.

Pasteur's report of 1860 provided the first example of the use of a synthetic medium for microbiological studies. During the fermentation, the yeast increased in weight with the consequent increase in the carbon and nitrogen contents of the synthetic medium was accompanied by a related decrease in the ammoniacal nitrogen of the medium, demonstrating that substrate nitrogen was being incorporated into the yeast cells.

Pasteur (1861) revealed that the growth and physiology of yeast (or other microorganism) differs depending on whether it is grown under aerobic or anaerobic conditions, which is now known as "Pasteur effect". Under anaerobic conditions relatively less growth occurred and that a large amount of sugar was converted to alcohol. He reasoned that the yeast was obtaining its required oxygen from the sugar molecule. Under aerobic incubation conditions less sugar was decomposed, and a greater portion of the sugar was converted to yeast cellular materials.

The famous death knell to spontaneous generation came in when Pasteur showed that microbes could occur as air contaminants. The he proceeded to sterilze media in a "Swan necked flask" so as to show that growth could not occur in sterile media unless air borne contamination had occured. Demonstrated the sterilization and maintenance of asepsis of fermentation media; the partial sterilization of media, or pasteurization, came at a later date. He reasoned that, if sequential dilutions of the culture were prepared, the contaminating microorganism probably would not be present in some of the greater dilutions which would still contain the microorganism of interest.

Ferdinand Cohn felt that perhaps certain microorganisms might exist which possess greater heat resistance than did the organisms previously studied. **John Tyndall** described "Intermittent sterilization" or Tyndalisation or "Fractional sterilization". The initial heat treatment causes "heat-shocking", with consequent germination of the spores during incubation, to yield the less resistant vegetative forms which are then killed by the next heat treatment.

Robert Koch (1881) employed gelatin as a solidifying agent and as a result was able to obtain individual colonies well separated from each other on the surface of the gelatin medium. Could be utilized for the description and identification of microorganisms, and that this technique also could be used for the purification of cultures. **Fanny Hesse**, the wife of one of Koch's students, realized that agar provided a solidifying agent that demonstrated none of the faults of gelatin or of solidified blood serum.

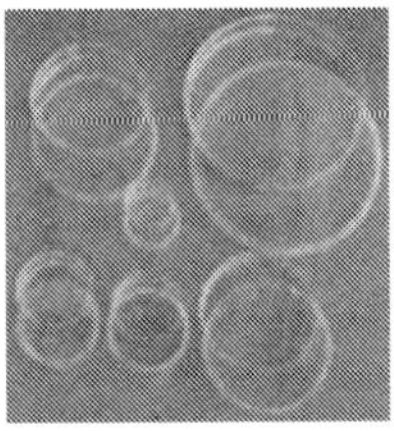

Petri plates

There was still the problem of a suitable culture vessel. In 1887, **Petri** solved this problem by describing what is now known as the Petri plate, a culture dish that is still used extensively today. **Eduard Buchner** (1897) ground up yeast cells and expressed the juice from the broken cells. He then filered the juice to obtain a clear solution, and mixed the solution with a concentrated sugar solution. The cell-free yeast juice carried out the alcoholic fermentation. He considered that the active agent was a soluble proteinaceous enzyme, called this enzyme "zymase" was extracellular and operated in the medium outside the yeast cells. Buchner was the first to demonstrate enzymatically mediated fermentation reactions. It is now known that zymase is not a single enzyme, but a series of enzymes making up the Meyerhof-Embden scheme for sugar catabolism, and that these enzymes are not normally extracellular but function within the yeast cell.

More than a minimal inoculum level was required to obtain growth of microorganisms in synthetic media. **Wildiers** (1901) demonstrating that yeast has a vitamin requirement for growth. Named this growth factor, or vitamin as "bios". He obtained the bios factor by extracting boiled yeast and, when this extract was utilized in conjunction with a low-level yeast inoculum, growth ensued in synthetic media.

F.d'Herelle (1917) coined the name "bacteriophage" for the filterable parasite of the dysentery bacillus that he was studying. He determined that the bacteriophage did not multiply in the absence of its living host. He also determined the host specificity range for this phage, and predicted that phages might exist for other microorganisms. His primary hope for bacteriophage that they could be used medicinally to treat disease or to confer immunity has not been realized.

Kluyver in 1924 pointed out, however, that regardless of all this diversity, there was an underlying unity among microorganisms (and higher forms of life) in both the metabolic pathways utilized and the associated energy release and utilization. He based these conclusions, in part, on observations that, by adjusting the growth medium or cultural conditions, one could change the types of metabolic intermediates (or end products) which accumulated during the growth of various microorganisms.

Alexander Fleming accidentally identified the penillin producing fungi *Penicillium notatum.* He employed streak plates to determine the bacterial inhibition spectrum for penicillin. Fleming also employed the dilution tube assay to determine the bacterial inhibition spectrum, as well as to assay the potencies of various penicillin preparations. Later penicillin was commercially produced from *Penicillium chrysogenum.*

It should now be obvious, from this brief history, that many of the earlier researchers did not set out to solve a problem or to interpret their results in terms of industrial microbiology applications. However, it is apparent that many of these studies do have a strong bearing on the technology and thinking of modern-day industrial microbiology, as well as on the technology of the times in which the studies were carried out.

❑❑❑

Chapter 22

Types of Fermentation Processes

The fermentation unit in industrial microbiology is analogous to a chemical plant in the chemical industry. A fermentation process is a biological process and, therefore, has requirements of sterility and use of cellular enzymic reactions instead of chemical reactions aided by inanimate catalysts, sometimes operating at elevated temperature and pressure. The industrial fermentations are mainly done in 3 ways *viz.*, surface fermentation, submerged fermentation and solid state fermentation.

I. Types of fermentation based on growth of producer organism

a. Surface fermentation

In this fermentation the producer organism is grown on the surface of the fermentation broth. This type is mainly used for those industrial processes that involve aerobic microbes.

b. Submerged fermentation

In this fermentation the producer organism is grown in submerged state in the fermentation broth. This type is mainly used for those industrial processes that involve anaerobes and facultative anaerobes. Most of the industrial processes are performed in this way.

c. Solid state fermentation

Solid state fermentations (SSF's) involves the growth of microorganisms on moist solid substrate particles in the absence or mere absence of visible liquid water between the particles. The moisture content of solid substrate ranges between 12 – 80 %, although typical water content is near the middle of this range. But the water content of a typical submerged fermentation is more than 95% and appears similar to water.

SSF's are usually used for the fermentation of agricultural products or foods, such as rice, wheat, barley, corn and soybeans. The low moisture levels of SSF's acts as a powerful selection pressure for the growth of mycelial organisms.

Some food fermentations involving SSF

Soysauce from soybean or wheat by *Aspergillus*

Tempeh from soybean by *Rhizopus*

Miso from rice / Soybean by *Aspergillus*

Phases of SSF

There are 3 phases in Solid State Fermentation

i. *Solid phase* : This can be of 2 types.
   a. Natural solid materials such as food stuff, agricultural products or by products, etc. The materials act as the main source of nutrients and the organism releases enzymes for the break down and modification of solid materials. Eg. Cereal straw, bran, soybean, begasse
   b. An inert solid support supplemented with nutrients for microbial growth. Act as physical support for growth through which nutrients and enzymes released by microbes can diffuse. Eg. Polystyrene, Styrofoam

ii. *Liquid phase* : This is the very thin liquid layer bound to the solid phase where intraparticle mass transfer occurs.

iii. *Gaseous phase* : This includes the void spaces which provides oxygen for growth of microbes. This can be manipulated for temperature and moisture conditions in SSF.

Characteristics of solid state fermentations

- The substrate is usually of agricultural origin and might be grains such as rice, tubers such as cassava, processed materials such as corn meal or byproducts such as straw
- The substrate may require preparation or pretreatment like,

  (a) Chopping or grinding is done to reduce particle size (b) Cooking or chemical hydrolysis is done to increase susceptibility of macromolecules to degradation (c) Pasteurization or sterilization is done to reduce contaminants.
- Microorganism is usually a filamentous fungus requiring aerobic condition.
- The inoculum is mixed into substrate to initiate the fermentation
- The bioreactor may or may not allow for mixing and aeration
- On completion of the process, the whole fermented substrate is taken out in case of food or feed. But other metabolites must be recovered through downstream processing.

Advantages of SSF's over conventional submerged fermentations are:

1. The small volume of fermentation mash or reactor volume results in lower capital and operating costs
2. A lower chance of contamination due to low moisture levels
3. Ease of product separation
4. Energy efficiency
5. Allows the development of fully differentiated structures, which is critical in some cases to product formation

Disadvantages

1. Heterogeneous nature of the media, due to poor mixing characteristics results in control problems (pH, DO, temperature) within the fermentation mash
2. For large fermentation mash volumes, it can be difficult to provide sufficient mixing to prevent concentration gradients from forming

3. At high agitation speeds, mycelial cells may be damaged

   (Rotary-tray or rotating-drum fermenters are often used to provide gentle and adequate agitation in SSF's).

II. Types of fermentation based on duration and Recovery method

Based on the duration of fermentation, production capacity and recovery method the fermentation methods are further divided into 3 types. They are

- Batch
- Continuous
- Fed Batch

a. Batch Fermentation

Batch fermentation is a discontinuous process wherein nutrient addition and product recovery is done in batches. The fermenter has to be cleaned after each process and a fresh batch started. It is considered to be a *"closed system"*. It includes the following 5 steps.

1. Medium added
2. Fermentor sterilised
3. Inoculum added
4. Fermentation followed to completion
5. Culture harvested

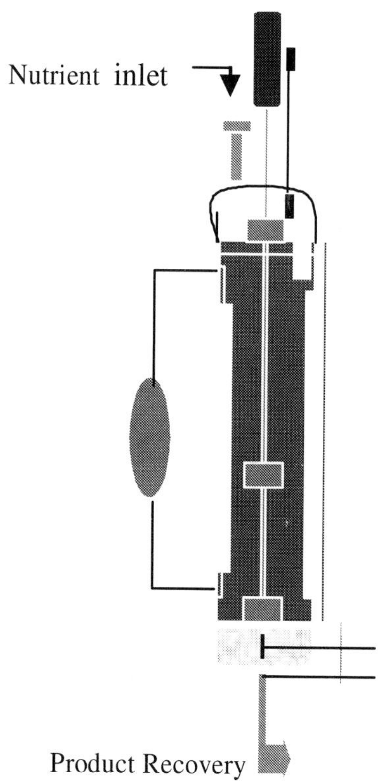

In this the growth of the culture follows a characteristic growth curve with a lag phase followed by log phase finally reaching the stationary phase and death phase due to limitation of nutrient and other factors. As the growth increases there is increase in cell biomass and product formed. This is the time to stop fermentation and harvest the product.

Characteristics of a Batch Fermentation System

- Simplest fermentor operation
- Sterilisation can be performed in the reactor
- All nutrients are added before inoculation
- Maximum levels of C and N are limited by inhibition of cell growth
- Biomass production limited by C/N load and production of toxic waste products

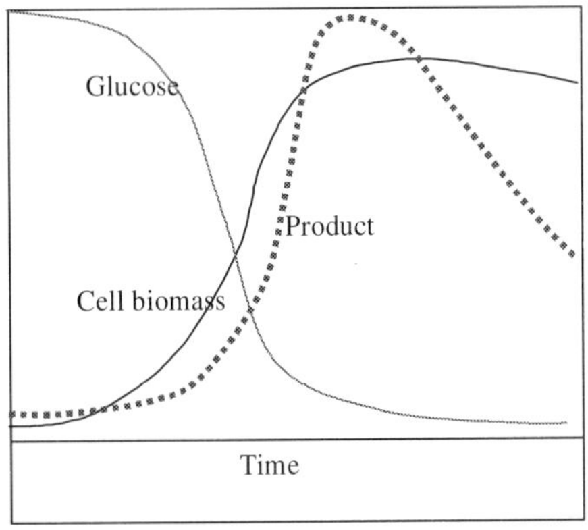

- Cells are harvested when biomass levels or product levels start to decline

b. Continuous fermentation

It is a continuous process where the nutrient is continuously added to the fermenter at a fixed rate, the organisms are continuously maintained at logarithmic stage and the products are recovered continuously. The fermenters in this type are called *"flow through"* fermenters.

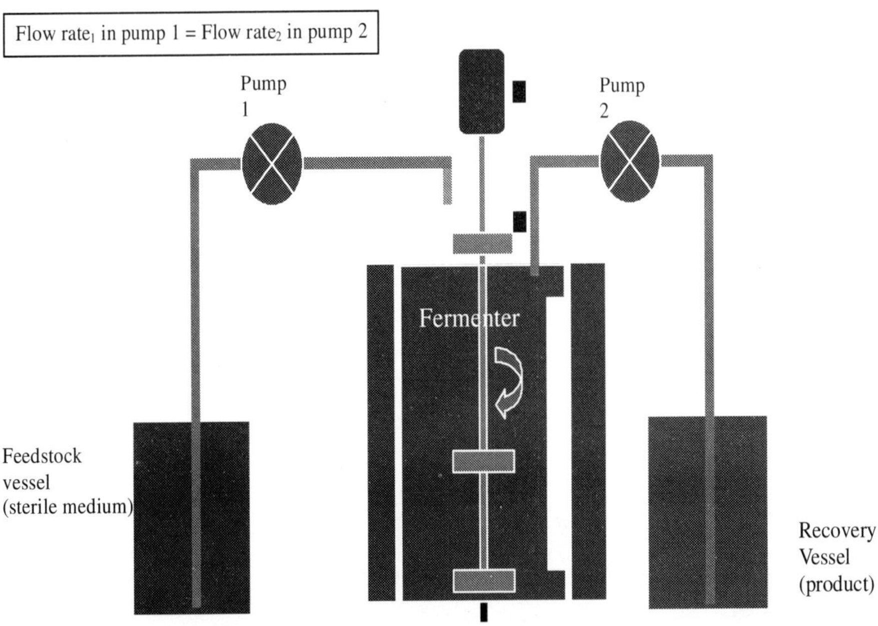

Continuous culture can *outperform batch* culture by *eliminating* the inherent down *time for cleaning and sterilization and the long lags* before the organisms enter a brief period of high productivity. The graph shows typical batch operation where the fermenter is used over and over again.

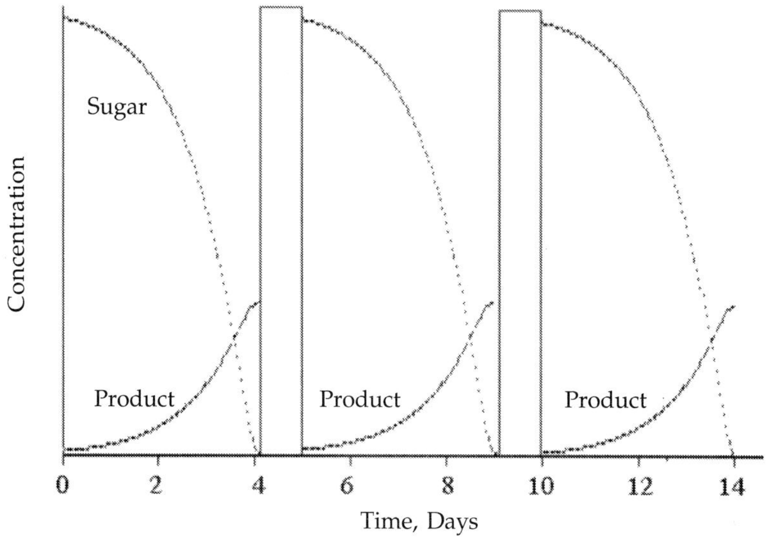

Graph showing the **inherent down time** for cleaning and sterilization and the long lags before the organisms enter a brief period of high productivity in **batch fermentation**

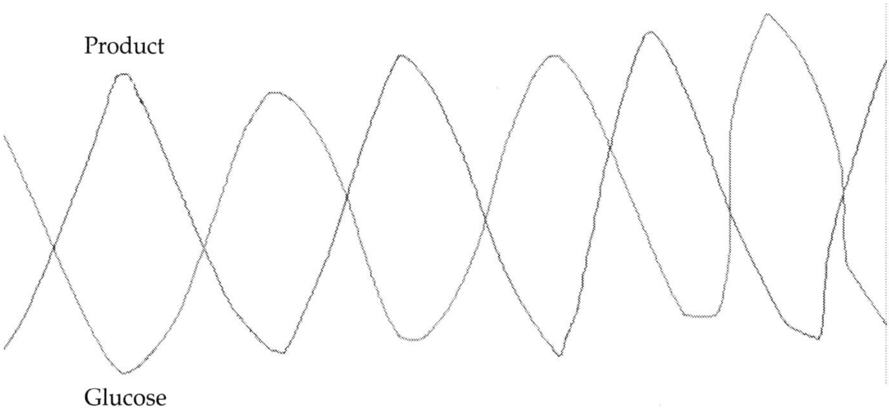

The level of substrate/ Nutrient and product in due course of **Continuous fermentation**

Advantages

1. Continuous fermentation allows better vessel and space utilization through faster fermentation.
2. The fermentation proceeds more rapidly because it is stirred and higher cell concentrations are present than in standard batch fermentations
3. Capital and labor costs are reduced since there is only one fermentation to control and test.
4. The fermentation system can be tuned according to market demands; in peak periods the CF can be run at a high flow rate, during the off-peak season the flow rates are reduced so that product is produced more slowly by the system.
5. A number of parameters (such as culture concentration, oxygenation rate and flow rate) can be used to fine-tune the quality of the product produced from the system.
6. The control of the fermentation also has the advantage of being automated.

Disadvantages of Continuous Fermentation

1. Hygiene requirements are much higher than those required for batch fermentation since a CF system has the potential to run for a year or more during which time the vessels cannot be cleaned.
2. More prone to contamination. Once contaminated the whole set up has to be dismantled and sterilized thoroughly.

c. Fedbatch Fermentation

This fermentation is intermediate of both batch and continuous fermentations. Nutrients are added continuously in a particular flow rate as in continuous fermentation and the product is recovered in batch after some time period. A proper feed rate, with the right component constitution is required during the process.

Characters of Fed batch fermentation

- Initial medium concentration is relatively low (no inhibition of culture growth)
- Medium constituents (concentrated C and/or N feeds) are added continuously or in increments (as in continuous fermentation)

Pump

Feedstock
vessel
(sterile)

- Controlled feed results in higher biomass and product yields
- Fermentation is still limited by accumulation of toxic end products
- Finally the products are harvested in one stroke (as in batch fermentation)

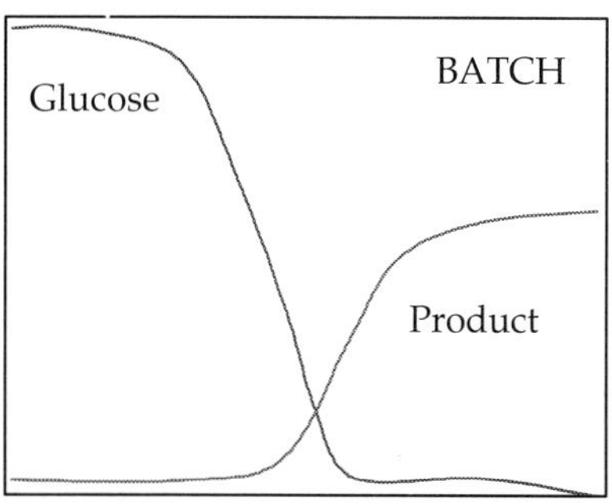

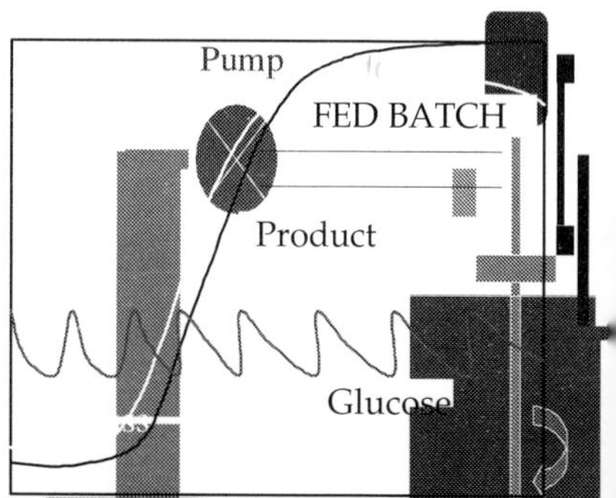

Advantages

1. Production of high cell densities due to extension of working time
2. Controlled conditions in the provision of substrates during the fermentation, particularly regarding the concentration of specific substrates as for ex. the carbon source
3. Control over the production of by-products or catabolite repression effects due to limited provision of substrates solely required for product formation

4. The mode of operation can overcome and control deviations in the organism's growth pattern as found in batch fermentation
5. Allows the replacement of water loss by evaporation
6. Alternative mode of operation for fermentations leading with toxic substrates or low solubility compounds
7. Increase of antibiotic-marked plasmid stability by providing the correspondent antibiotic during the time span of the fermentation
8. No additional special piece of equipment is required as compared with the batch fermentation mode of operation.

Disadvantages

1. It requires previous analysis of the microorganism, its requirements and the understanding of its physiology with the productivity.
2. It requires a substantial amount of operator skill for the set-up, definition and development of the process.
3. In a cyclic fed-batch culture, care should be taken in the design of the process to ensure that toxins do not accumulate to inhibitory levels and that nutrients other than those incorporated into the feed medium become limiting.
4. If many cycles are run, the accumulation of non-producing or low-producing variants may result.
5. The quantities of the components to control must be above the detection limits of the available measuring equipment.

□□□

Chapter 23

Stages of Industrial Processes

The producer organism, fermenter, fermentation process, product and recovery are the important components of Industrial fermentation. These components are interlinked in many industrial processes.

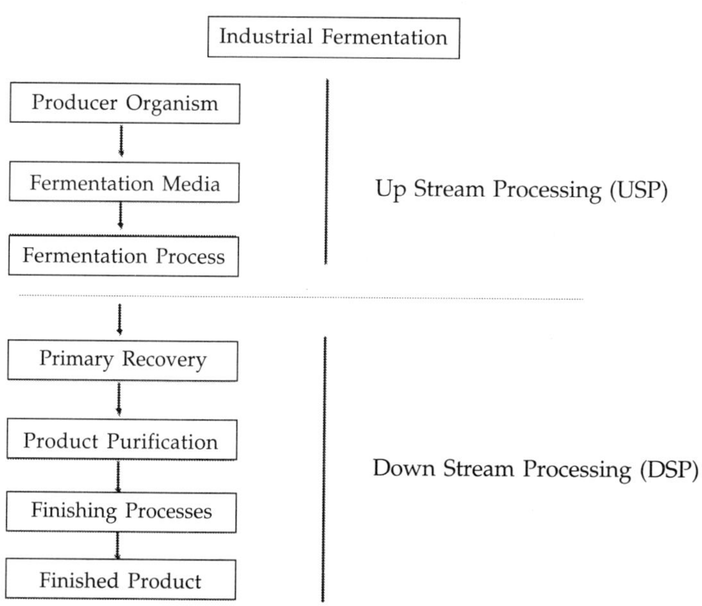

Industrial fermentation includes 2 stages of processes viz., Up stream Processing (UPS) and Down Stream Processing (DSP).

I. Up Stream Processing : USP includes all factors and processes leading to fermentation which consisits of 3 main areas.

Area 1: Aspects associated with Producer organism. This includes the following stages

a) strategies for obtaining a suitable microorganism
b) industrial strain improvement for enhancing productivity and yield
c) maintenance of strain purity
d) preparation of a suitable inoculum
e) continuing development of selected strains to increase the economic efficiency of the process

Area 2: Fermentation media- selection of suitable cost effective carbon and energy sources along with other essential nutrients with a view to maximize the yield

Area 3: Fermentation process – conditions for the optimum growth of microbes or for the development of microbial product

II. Down Stream Processing : Downstream processing refers to the recovery and purification of biosynthetic products from fermentation broth. DSP involves all the other processes after fermentation. It includes primary recovery, product purification, finishing processes and finished product.

We will discuss about each step of the USP and DSP in this chapter.

Selection and Screening of Producer Microbes

Metabolites are the intermediates and products of metabolism. Metabolites from chemical compounds, whether inherent or pharmaceutical, are formed as part of the natural biochemical process of degrading and eliminating the compounds. The rate of degradation of a compound is an important determinant of the duration and intensity of its action. They are of 2 types, primary and secondary metabolite.

Primary metabolite: These are the metabolites produced during the growth phase of the microbe and directly involved in normal growth, development, and reproduction. Examples: alcohol, amino acids, nucleotides, organic acid, vitamins and enzymes.

Industrially produced primary metabolites:

Class	Example
Alcohol	Ethanol
Amino Acids	Glutamic Acid, Aspartic acid
Nucleotides	5' guanylic acid
Antioxidants	Ascorbic acid, Glutathione
Organic Acids	Acetic Acid, Lactic Acid
Polyols	Glycerol
Vitamins	B_2, B_{12}

Secondary metabolites : These are the organic compounds that accumulate during periods of nutrient limitation and waste buildup but are not directly involved in the normal growth, development, or reproduction of organisms. Unlike primary metabolites, absence of secondary metabolities does not result in immediate death, but rather in long-term impairment of the organism's survivability, fecundity, or aesthetics, or perhaps in no significant change at all. Secondary metabolites often play an important role in defense system. Example: antibiotics, pigments, toxins. Some secondary metabolites use primary metabolites as precursors, such as actinomycin which is synthesised from the primary metabolite, tryptophan.

Industrially produced secondary metabolites

Class	Example
Antibiotics	Penicillin, Streptomycin, Tetracyclin, Grieseofulvin, Cephalosporin
Pigments	Zeaxanthin, Astaxanthin, β-carotene, lycopene, carotenoids

Characteristics of an ideal Producer Microbe

The basic foundation of any industrial fermentation is laid by the efficiency of the microbes used. The success of an industrial fermentation process chiefly depends on the microorganism strain used. An ideal producer or economically important strain should have the following characteristics.

1. It should be pure, and free from phage.
2. It should be genetically stable, but amenable to genetic modification by mutagenesis.
3. It should grow vigorously after inoculation in fermentation vessels.
4. It should produce a valuable product, and no toxic by-products.
5. The product should be produced in a short time, e.g., 2-3 days.
6. It should be amenable to long-term conservation.
7. The risk of contamination should be minimal under the optimum performance conditions.

As ideals are rarely achieved, most industrially valuable microbial strains are usually a compromise among the above features.

❒❒❒

Chapter 24

Industrial Production of Pharmaceuticals

Fermentation technology has contributed to the production of various chemicals such as antibiotics, vitamins, steroids and vaccines, thus being the mainstay of pharmaceutical industry. Antibiotics production itself is a major industry. Some of the important industrial processes of the production of pharmaceuticals are discussed here.

Antibiotics

Antibiotics are the chemical substance produced by microorganisms; that inhibit or kill the growth of other microorganism even at very low concentrations, it is a product of secondary metabolism. The significance of antibiotic production for the producing strain is unclear. As secondary metabolite, antibiotics could serve regulatory roles during differentiation, perhaps acting as temporary inhibition agents.

Alexander Flemming (1929) was the first to report the inhibitory effect of the mold *Penicillium notatum* on the growth of *Staphylococcus aureus*.

Development of antibiotics as agents for treatment of infectious diseases had more impact on the practice of medicine. Large numbers of known antibiotics are microbial origin and major antibiotics are produced by fermentations and semisynthetic processes. Only very few are produced synthetically.

Microbial groups producing antibiotics

Antibiotics are produced by bacteria, actinomycetes and fungi. In fungi, the antibiotics produced by Aspergillaceae and Moniliaceae are of practical importance. Many bacteria are producing antibiotics. Greatest variety of antibiotics are produced by actinomycetes, especially the genus *Streptomyces*. Two third of total antibiotics are produced by actinomycetes only.

Classification of antibiotics

Antibiotics are classified according to their

- Antimicrobial spectrum
- Mechanism of action
- Producer strain
- Manner of biosynthesis
- Chemical structure

a. Based on antimicrobial spectrum

Based on the ability of the antibiotic to inhibit Gram positive bacteria, or Gram negative bacteria, or against Fungi, or Broad spectrum the antibiotics are calssified

Antibacterial antibiotics	:	Penicillin, Streptomycin
Antifungal antibiotic	:	Cycloheximide, Nystatin
Broad spectrum antibiotic	:	Tetracycline, Chloromphenicol

b. Based on mechanism of action

Antibiotics are classified based on their action on cell wall or cytoplasmic membrane or inhibiting protein or nucleic acid synthesis

Inhibit the Cell wall synthesis	:	Penecillin,Bacitracin
Inhibit the Protein synthesis	:	Streptomycin, Tetra cycline, Cycloheximide
Inhibit the Cell membrane synthesis	:	Nystatin

c. Based on Producer strain

Based on the origin of the antibiotic, it may be classified as Antibiotics of bacterial origin/fungal origin/actimycetes/from plant origin

Bacteria origin : Bacitracin, polymyxin
Fungal origin : Penicillin, Cephalosporin
Actinomycets origin : Streptomycin, Tetracycline

d. Based on the manner of biosynthesis

Natural antibiotics - Normal fermentation yield natural antibiotics.

Biosynthetic antibiotics - Production of a particular form of antibiotic by supplying the organisms with specific side chain precursors

Semi synthetic antibiotics - produced chemically by adding a specific side chain to the antibiotics

Synthetic antibiotics - produced fully by chemical process.

e. Based on chemical structure antibiotics are grouped as follows

- Carbohydrate containing- Steptomycin, Vancomycin.
- Macrocyclic lactones - Erythromycin , Rifampicin.
- Quinones & related antibiootics - Tetracycline, Adiramycin
- Amino acid and peptide antibiootics- Penicillins, Bacitracin
- Heterocyclic antibiotics containing N- Polyoxins
- Heterocyclic antibiotics containing oxygen- Polyether Antibiotics
- Alicyclic derivatives – cyclohexamide
- Aromatic antibiotics – Chloramphenicol, Griseofulvin.
- Aliphatic anitibiotics – Fosfomycins.

1. Penicillin

Penicillin's ability to cure people of many once-fatal bacterial infections has saved so many lives that it is easy to understand why it was once called a *"miracle drug"*.

Norman Heatley, whose technical innovations at the Sir William Dunn School of Pathology, Oxford University, made possible the large scale production of penicillin for its first tests on humans in 1941.

Commercially produced antibiotics and their producers

Antibiotic	Producing microorganism
Penicillin	*Penicillium chrysogenum*
Cephalosporin	*Cephalosporium acremonium*
Grieseofulvin	*Penicillium griseofulvum*
Chloramphenicol	*Sterptomyces venezuelae*
Tetracyclines	
Tetracycline	*Sterptomyces aureofaciens*
Chlortetracycline	*S. aureofaciens*
Oxytetracycline	*S. rimosus*
Polypeoptides	
Polymyxin – B	*Bacillus polymyxa*
Bacitracin	*B. licheniformis*
Glutarimides	
Cycloheximide	*Streptomyces griseus*
Aminoglycosides	
Streptomycin	*Streptomyces griseus*
Kanamycin	*S. kanamyceticus*
Neomycin	*S. fradiae*
Polyenes	
Nystatin	*Sterptomyces noursei*
Aureofungin	*Streptomyces sp.*
Macrolides	
Erythromycin	*Streptomyces erythreus*
Blasticidin-S	*Streptomyces sp.*
Vira –A (Adenine arabinoside)	*Streptomyces antibioticus*

In 1929, **Alexander Fleming**, a doctor and researcher at St. Mary's Hospital in London, England, published a paper on a chemical called "penicillin", which he had isolated from a mold, ***Penicillium notatum***. Fleming wrote, that penicillin had prevented the growth of a neighboring colony of germs in the same petri dish but he was never able to purify his samples of penicillin, but he became the first person to publish the news of its germ-killing. Howard Florey and Norman Heatley developed it further for enough production of penicillin.

Alexander Fleming

Howard Florey

Norman Heatley

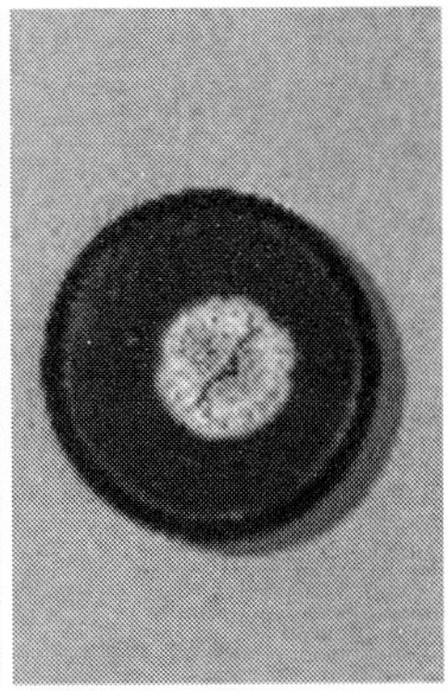

Petri dish containing colony of *Penicillium notatum*. Sold by Christie's Auction House in 1998 for $13,121.

Production of Penicillin

A selected high yielding strain of *Penicillium chrysogenum* is used for penicillin production.

Culture	:	*Penicillium chrysogenum*. The inoculum is raised in wheat bran nutrient solution by allowing the culture to incubate for 1 week at 24°C. The cultures raised in flasks are transferred to inoculum tanks and aerated for 1-2 days to produce more inoculum with heavy mycelial growth.
Medium	:	10 per cent total glucose or molasses, 4-5 per cent cornsteep liquor solids, 0.5-0.8 per cent total phenylacetic acid and 0.5 per cent total vegetable oil, by continuous feed. Phenylacetic acid is the precursor used for synthesizing the benzene ring side chain of Penicillin-G. The addition of the precursor stimulates the fungus to produce more Penicillin-G.
Incubation Temperature	:	25-26°C
pH of medium	:	6.0
Oxygen supply	:	Aerobic
Fermentation period	:	7 days

The production is carried out in huge production tanks which are well aerated and the fungus grows as a submerged culture mostly as hyphal balls. In earlier methods, lactose was being added to the medium but now this is totally eliminated. The fermentation is allowed to go on for 7 days. During the first day of fermentation, mycelial growth takes place with the utilization of carbohydrates. Reduction in the carbohydrate concentration leads to a condition favouring penicillin production from the second day and the penicillin production continues from second day till the seventh day. The pH rises to 8.0 as the fermentation goes on and penicillin production stops at this stage.

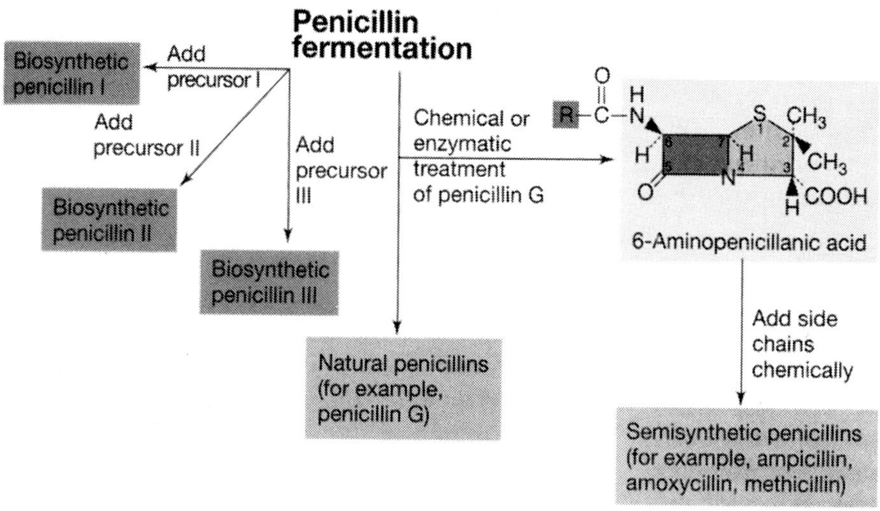

Harvest and Recovery of Penicillin

- Harvest of culture broth using rotary vacuum filter
- Remove mycelium and solids
- Acidification of broth using phosphoric or sulfuric acid – pH 2.0 to 2.5
- Extraction with organic solvent - amyl acetate
- Back extraction in water by adding potassium/sodium salt
- Washing and drying of crystalline Penicillin
- Spent solvents are to be recovered for reuse
- Various other modifications of Penicillin-G are carried out chemically and biologically

2. Cephalosporins

Cephalosporin-C is made from the fermentation by *Cephalosprium acremonium*, the process being almost similar to Penicillin production. However, Cephalosporin-C is not quite potent and needs to be chemically converted into 7-alphaaminocephalosporanic acid to form clinically useful products.

3. Streptomycin

Culture	:	*Streptomyces griseus*. The spore inoculum of *Streptomyces griseus* is raised just as in penicillin production and transferred to fermentation tanks
Medium	:	Soybean meal, glucose and sodium chloride
Incubation Temperature	:	30°C
pH of medium	:	7.6-8.0
Oxygen supply	:	Aerobic
Fermentation period	:	10 days

Commercial fermentation involves three steps

First phase

- First 24 hours, rapid growth occurs and during this period most of the mycelium for the fermentation are produced.
- The strong proteolytic activity of *Streptomyces griseus* releases ammonia to the medium from the soybean meal.Carbon nutrients of the soybean meal are utilized for growth.
- The glucose of the medium is utilized only slowly during this period and only slight streptomycin production occurs.

Second phase

- Streptomycin produced at a rapid rate, lasts from approximately 24 hours to 6 or 7 days of incubation.
- No mycelial growth occurs, weight of the mycelium remains fairly constant. Ammonia is utilized and the glucose is rapidly withdrawn from the medium.

Third phase

- The sugar is depleted from the medium and streptomycin production ceases
- The cells lyse, releasing ammonia and the pH rises

Harvest and recovery

- Broth filtration to remove mycelium
- The streptomycin is adsorbed from the broth onto activated carbon and then eluted from the carbon with dilute acid
- The antibiotic is then precipitated with solvent (acetone) and further purified using column chromatography.

Vitamin B_{12} as by product during streptomycin production

- Small amount of vitamin B_{12} in addition to streptomycin without affecting streptomycin yields (soluble cobalt salt is added as a precursor) is also present.
- Vitamin B_{12} pure form from the culture broth or dried fermentation residues or partially purified concentrate of the vitamin added as a supplement to animal feeds.
- Commercial Vit B_{12} is by microbial fermentation.

❑❑❑

Chapter 25

Industrial Production of Organic Acids

Several organic acids such as lactic acid, citric acid, acetic acid, gluconic acid,itaconic acid and gibberellic acid are being produced by microbial fermentation. Some of the production processes are briefty described here.

1. Acetic Acid

Acetic acid (also called vinegar) is produced both synthetically and by bacterial fermentation. Today, the biological route accounts for only about 10% of world production, but it remains important for vinegar production, as many of the world food purity laws stipulate that vinegar used in foods must be of biological origin.The global demand of acetic acid is around 6.5 million tonnes per year (Mt/a), of which approximately 1.5 Mt/a is met by recycling; the remainder is manufactured from petrochemical feedstocks or from biological sources.

Fermentation

Oxidative fermentation

A dilute alcohol solution inoculated with *Acetobacter* and kept in a warm, airy place will become acetic acid over the course of a few months. Industrial aceticacid production methods accelerate this process by improving the supply of oxygen to the bacteria. Commonly used feeds include apple cider, wine, and fermented grain, malt, rice, or potato mashes. The overall chemical reaction facilitated by these bacteria is

$$C_2H_5OH + O_2 \rightarrow CH_3COOH + H_2O$$

Types of acetic acid Production

- Open vat method
- Trickle method
- Bubble method
- Anaerobic fermentation

a. Fast method or German method or Vat method

One of the first modern commercial processes was the "Vat method" or "fast method" or "German method", first practised in Germany in 1823.

In this process, fermentation takes place in a tower packed with wood shavings or charcoal. The alcohol-containing feed is trickled into the top of the tower, and fresh air supplied from the bottom by either natural or forced convection. The improved air supply in this process cut the time to prepare vinegar from months to weeks.

Vat method

b. Trickle method

Trickling of Alcoholic liquid over wood shaving packed in a vat (or) column while a stream of air enters at the bottom and passes upward.

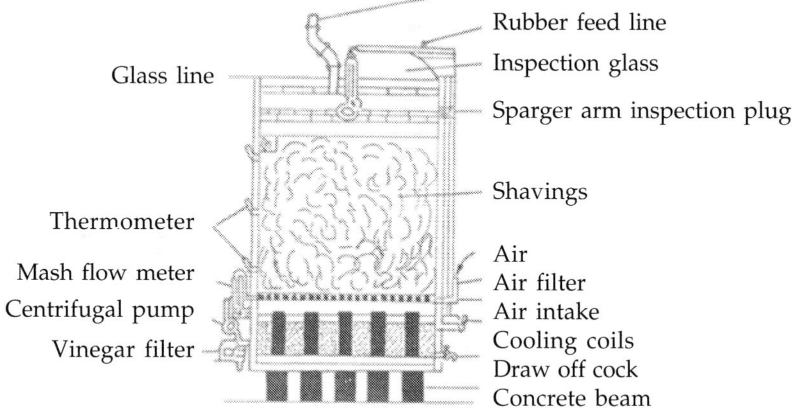

c. Bubble Method

The bacteria is grown under submerged condition in the broth and this method has 90-98% efficiency. Most vinegar today is made in submerged tank culture, first described in 1949 by *Otto Hromatka* and *Heinrich Ebner*. In this method, alcohol is fermented to vinegar in a continuously stirred tank, and oxygen is supplied by bubbling air through the solution. Using this method, vinegar of 15% acetic acid can be prepared in only 2–3 days.

d. Anaerobic fermentation

Some species of anaerobic bacteria, including several members of the genus *Clostridium*, can convert sugars to acetic acid directly, without using ethanol as an intermediate. The overall chemical reaction conducted by these bacteria may be represented as:

$$C_6H_{12}O_6 \rightarrow 3\ CH_3COOH$$

More interestingly from the point of view of an industrial chemist, many of these acetogenic bacteria can produce acetic acid from one-carbon compounds, including methanol, carbon monoxide, or a mixture of carbon dioxide and hydrogen:

$$2\ CO_2 + 4\ H_2 \rightarrow CH_3COOH + 2\ H_2O$$

This ability of *Clostridium* to utilise sugars directly, or to produce acetic acid from less costly inputs, means that these bacteria could potentially produce acetic acid more efficiently than ethanol-oxidisers like *Acetobacter*. However, *Clostridium* bacteria are less acid-tolerant than *Acetobacter*. Even the most acid-tolerant *Clostridium* strains can produce vinegar of only a few per cent acetic acid, compared to some *Acetobacter* strains that can produce vinegar of up to 20% acetic acid. At present, it remains more cost-effective to produce vinegar using *Acetobacter* than to produce it using *Clostridium* and then concentrating it.

Applications

1. It is an important chemical reagent and industrial chemical that is used in the production of polyethylene terephthalate mainly used in soft drink bottles;
2. cellulose acetate, mainly for photographic film; and polyvinyl acetate for wood glue, paints, adhesives, as well as many synthetic fibres and fabrics.

3. In households diluted acetic acid is often used as descaling agents to remove limescale from taps and kettles.
4. In the food industry acetic acid is used as an acidity regulator and as a food preservative (Sodium acetate).
5. Acetic anhydride is also a reagent for the production of aspirin, heroin, and other compounds.
6. In the form of vinegar, acetic acid solutions (typically 5% to 18% acetic acid, with the percentage usually calculated by mass) are used directly as a condiment, and also in the pickling of vegetables and other foodstuffs.
7. Used as solvent in recrystallisation of organic compounds for purification and commercial manufacture of synthetic camphor
8. Copper(II) acetate—used as a pigment and a fungicide.
9. The acidity is also used for treating the sting of the box jellyfish by disabling the stinging cells of the jellyfish.
10. As a spray-on preservative for livestock silage, to discourage bacterial and fungal growth.

2. Lactic Acid

The lactic acid bacteria belong to two main groups – the homofermenters and the heterofermenters. The pathways of lactic acid production differ for the two. Homofermenters produce mainly lactic acid, via the glycolytic (Embden–Meyerhof) pathway). Heterofermenters produce lactic acid plus appreciable amounts of ethanol, acetate and carbon dioxide, via the 6-phosphoglucanate/ phosphoketolase pathway. Normal conditions required for this pathway are excess sugar and limited oxygen.

Homolactic fermentation

The fermentation of 1 mole of glucose yields two moles of lactic acid;

$$\underset{\text{Glucose}}{C_6H_{12}O_6} \rightarrow \underset{\text{lactic acid}}{2\ CH_3CHOHCOOH}$$

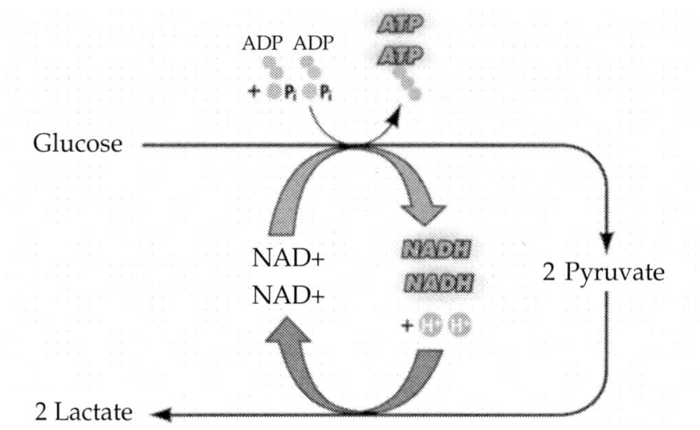

Heterolactic fermentation

The fermentation of 1 mole of glucose yields 1 mole each of lactic acid, ethanol and carbon dioxide;

$C_6H_{12}O_6 \rightarrow$	$CH_3CHOHCOOH +$	$C_2H_5OH +$	CO_2
Glucose	lactic acid+	ethanol+	carbon dioxide

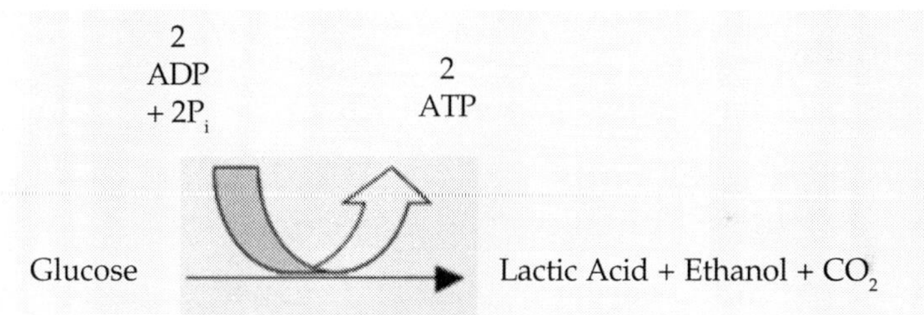

Industrial fermentation of lactic acid

Culture	:	*Lactobacillus delbrueckii* or *L. bulgaricus*
Medium	:	molasses, corn sugar, potato starch or whey as carbon source. Ammonium phosphate and trace amounts of other nitrogen sources are added. Ten per cent calcium carbonate is added to neutralize the lactic acid formed
Incubation Temperature	:	45-50°C
pH of medium	:	5.5-6.5
Oxygen supply	:	Anaerobic
Fermentation period	:	5-7 days

Lactobacillus bulgaricus

After fermentation, calcium carbonate is added to rise the pH to 10. The lactalbumin is coagulated by heating and filtering. Heating also inactivates bacteria. Calcium lactate is obtained in the process and it is crystallized. High purity of lactic acid is difficult and expensive to achieve. Pure lactic acid is produced only for analytical purposes.

Applications

Lactic acid has several commercial uses.

1. It is used in foods as a preservative and as an acidulant in confectionery, fruit juices and essences.
2. It is used in the curing of meat and canned vegetable products.
3. It is used in leather industry for deliming hides and in the textile industry for fabric treatment.
4. Various salts of lactic acid are used in industry, e.g., polylactic acid is used in resins, copper lactate is used in electroplating and calcium lactate is used in backing power and as an animal feed supplement.
5. Used in the production of fermented dairy products
6. Lactic acid is also used in labroratory stains such as lactophenol and lactofuchsin.

3. Citric Acid

Citric acid used to be extracted from citrus fruits but is now largely produced by microbial fermentation.

Industrial fermentation of citric acid

Culture	:	*Aspergillus niger* and *A. wentii*
Medium	:	Molasses, ammonium nitrate, magnesium sulphate and potassium phosphate.
Incubation Temperature	:	25-30°C
pH of medium	:	1.6 to 2.2
Oxygen supply	:	Aerobic
Fermentation period	:	5-7 days

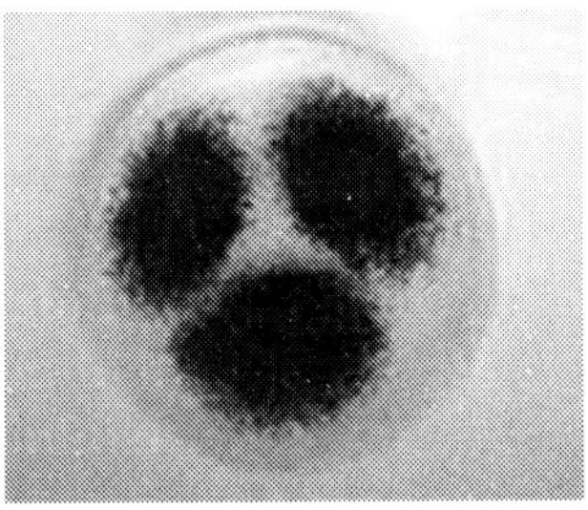

Aspergillus niger

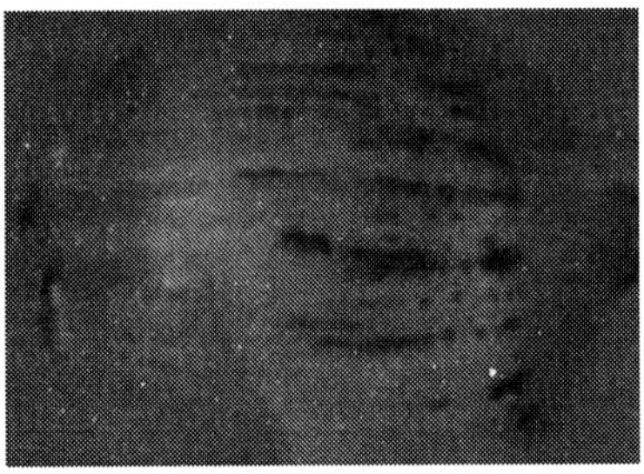

Aspergillus niger on onion

The sterile medium is dispensed in shallow pans and inoculated with mould spores to produce a surface culture. The fermentation is aerobic and more surface area provided, the better. The fungal mat is formed on the surface. The fermentation can also be carried out by submerged culture methods with continuous aeration. The citric acid is finally recovered as calcium citrate crystals.

Applications

Citric acid is commercially used in foods, soft drinks and certain pharmaceuticals.

It is also used in the manufacture of ink, dyes, etc.

Citric acid finds use also in leather tanning and electroplating.

4. Gluconic Acid

Normally submerged fermentation with artificial aeration is followed.

Industrial fermentation of Gluconic acid

Culture : *Aspergillus niger* (*Acetobacter* sp. and *Penicillium* spp. also produce gluconic acid.

		Since *A. niger* coverts glucose to gluconic acid in a single enzymatic reaction it is commercially used)
Medium	:	25 percent glucose, calcium carbonate and a compound containing element boron. Boron is added in the medium to prevent the precipitation of calcium gluconate
Temperature	:	30°C
pH of medium	:	6.5
Oxygen supply	:	Aerobic
Fermentation period	:	5 day

The fermentation is carried out in submerged cultures that are aerated and the temperature maintained at 30°C. After the growth of the *A. niger* mycelium is fully achieved in the initial 24-48 hours, the conversion of glucose to gluconic acid takes another 36 hours. Free gluconic acid accumulated is converted to calcium gluconate by the addition of calcium carbonate. Accumulation of free acid is thus avoided to keep the pH near neutral. The recovery of gluconic acid is made by addition of calcium hydroxide to the culture filtrate. Gluconic acid forms precipitate as calcium gluconate.

Applications

1. Gluconic acid is used commercially mainly in the pharmaceutical industry. It is used for giving calcium and iron to human body as calcium gluconate and ferrous gluconate respectively.
2. In the detergent industry, gluconic acid is used for the removal of encrustrations of calcium and magnesium in vessels, due to the use of hard waters.

5. Gibberellic Acid

Gibberellic acid and related chemicals called gibberellins are one of the major groups of growth promoting hormones which play essential roles in the growth and development of plants. The Japanese scientists in 1926 discovered that an ascomycetous fungus *Gibberella fujikuroi* (asexual stage: *Fusarium moniliforme*) produced plant growth promoting hormones which were later named as gibberellins. Gibberellins have been shown to be produced by other microorganisms, e.g., sporophores of *Agaricus bisporus, Phallus* sp., *Boletus* sp. and *Grifola* and a few species of bacteria including species of *Rhizobium*. However, the quantities of gibberellins produced by these organisms are much less compared to that produced by *G. fujikuroi*.

All gibberellins have the gibbane ring, even though the configurations may vary slightly. A large number of gibberellins are now being designated as GA_1, GA_2, GA_3 and so on. Gibberelline are now known to be present in plants as normal constituents in small quantities, being biosynthesized in stem tips and leaves.

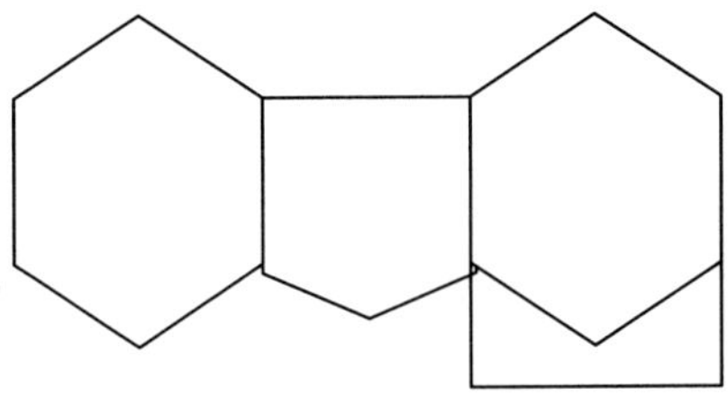

The Gibberellins

Industrial fermentation of Gibberellic acid

Culture	:	*Gibberella fujikuroi (= Fusarium moniliforme)* A type culture ACC 917 of *G. fujikuroi* has been successfully developed for production of GA_3, in the Regional Research Laboratory, Jammu, India.
Medium	:	Glucose (2%), $MgSO_4$ $7H_2O$(0.3%), NH_4CL (0.3%) cheaper substrates have been tried. For example, a combination of soybean flour and groundnut
Incubation Temperature	:	25- 30°C
pH of medium	:	slightly acidic – 6.0 to 6.5
Oxygen supply	:	aerobic
Fermentation period	:	7 days

It has been observed that 10-15 day old initial cultures incubated for 72 hrs in the first stage and 48 hrs in the second stage could produce maximum GA_3 in the production medium. The GA_3 from the medium is extracted with organic solvents and crystallized.

Applications

Gibberellins therefore find multifarious uses in horticulture.

They induce α-amylase activity in aleurone layers of seeds, produce a stimulus upon terminal meristems, control DNA dependent RNA synthesis, induce internode elongation, stimulate flowering and induce parthenocarpy.

❑❑❑

Chapter 26

Production of Microbial Enzymes

Enzymes have attracted the attention of scientists world over due to their wide range of physiological, analytical and industrial applications. Although enzymes have been isolated, purified and studied from microbial, animal and plant sources, microorganisms represent the most common source of enzymes due to their broad biochemical diversity, feasibility of mass culture and ease of genetic manipulation.

The industrial production of enzymes from microorganisms involves culturing the microorganisms in huge tanks where enzymes are secreted into the fermentation medium as metabolites of microbial activity. Enzymes thus produced are extracted, purified and used as processing aids in the food industry and for other applications.

Currently produced commercial enzymes

- Enzymes used in Industry – amylase, protease
- Enzymes used for analytical purpose – glucose oxidase, galactose oxidase
- Enzymes used in medicine – protease, lipase, streptokinase

Enzyme	Source	Application	Industry
Amylase (starch digesting)	Fungi,	Bread, Starch coatings, syrup and glucose manufacture, laundry starch,	Baking, paper, food, starch, pharmaceutical, textile, landury
	Bacteria	digestive aid, removal of coating (desizing), detergent	
Cellulase	Bacteria	fabric softening, brightening, detergent	Laundry
Protease (protein-digesting)	Yeast	Soft-center candies	Candy
Pectinase	Fungi	Pressing, clarification	Wine, fruit juice
Lipase	Fungi	Breaks down fat	Dairy, laundry
lactase	Fungi	Breaks down lactose to glucose and galactose	Dairy, health foods

Major microbial enzyme and their application

General Production methods

- *Semisolid culture* : The culture is grown in a surface of a solid substrate, usually moistened wheat or rice bran supplemented with nutrient salts. Sterilized medium is spread on the trays and inoculated with fungal spores and grown at aseptic condition
- *Submerged cultures* : Fermentation tank, equipped with agitator, an aerating device, cooling system is used. Production medium contain, energy source, C and N source and any growth requirements. Certain inducers are also needed.

1. Amylases

Amylases have various commercial applications. They are used in the textile and the paper industry mainly to dissolve and remove starch or to coat starch to paper. In laundry they find use in removal of spots from clothes in conjunction with proteases. Amylases also find use in the brewing industry in the initial breakdown of starchy substrates.

Amylases include α-amylase, β-amylases and glucamylases in addition to other categories, a-amylase converts starch to oligosaccharides and maltose; b-amylase converts starch to maltose and dextrins; glucamylase converts starch to glucose.

Microorganisms used

Fungi - *A.oryzae, Aspergillus niger, A. awamori , Rhizopus japonicus*

Bacteria - *Bacillus subtilis,Bacillus megaterium, B.polymyxa* and *B. diastaticus.*

Fermentation

Solid state/substrate fermentation

Submerged fermentation

Production of Fungal amylase

Solid state -Wheat-bran ,stationary culture has been employed extensively for the production of the fungal amylase. Submerged fermentation for amylase production have recently become economically feasible. *Aspergillus niger* is grown in submerged aerated culture using a starch-based medium to produce amylases.

Production of Bacterial amylase

Bacillus subtilis grown in a submerged aerated cultures. Bacterial amylase (has an optimum temperature for activity at approximately 55°C) are relatively heat resistant Eg: starch hydrolysis must be conducted at high temperature. The medium maintained at 25-30°C, pH -7.0 for 6days.

2. Proteases

Mixture of Proteinases and peptidases are called proteases. Proteases are excreted to the fermentation medium. Proteolytic enzymes also are employed in the textile industry to remove proteinaceous sizing, and silk industry to liberate the silk fiber, Tenderizing of meat, Dehairing in leather Industry, Active ingredient in spot remover, Removing food spots in the dry-cleaning industry.

The production of the enzyme is now carried out in submerged aerated cultures using a suitable medium, even though earlier methods used surface cultures. Usually the production takes a week depending on conditions provided.

Recombinant DNA technology has also led to the development of *Bacillus* strains that yield an enzyme called 'kerazyme' which dissolves hair and this enzyme is useful in clearing drains clogged with hair. Fungal proteases are used in the making of bread and are mainly derived from species of *Aspergillus*. Proteases are also used as digestive aids as they help digest meat products. Proteases find use in textile industry in removing proteinacious matter and in the leather industry for softening hides.

Microbes used

Bacteria : *Bacillus, Pseudomonas, Clostridium, Proteus* and *Serratia*

Fungi : *Aspergillus niger, A.oryzae,* A. *flavus* and *Penicillium roquefortii.*

Production of Fungal protease

Fungus is usually grown on wheat bran, as solid substrate fermentation although other media are sometimes employed (similar to those for amylase production)-Solid substrate. Submerged fermentation for fungal protease production may become commercially feasible.

Production of Bacterial protease

Submerged fermentation - The fermentation medium is incubated 3-5 days at 37°C, and harvested similar to that of bacterial amylase. Temperature should be less than 40°C in order to protect the enzyme from denaturation.

3. Pectic Enzymes

Pectin and pectic acid are present in the middle lamellae of plant tissues. Most phytopathogenic fungi and bacteria produce pectic enzymes as a means of macerating host plant tissues. The main groups of pectic enzymes produced by microorganisms are: (i) Pectin methyl esterases which remove the methyl group of pectin and convert it into pectic acid, (ii) polygalacturonases which cleave the α 1,4-glycosidic bonds of pectin or pectic acid by a hydrolytic mechanism and (iii) pectin lyases or pectin transeliminases which cleave the glycosidic bonds by a transeliminase mechanism without using water

Pectic enzymes are produced commercially from fungi such as *Aspergillus niger, Amillaria mellea* and bacteria such as *Erwinia amylovora*. Pectic enzymes find greatest use in apple juice industry where the enzymes help release maximum amount of juice by macerating the tissues.

Applications

Enzymes have applications in many fields, including organic synthesis, clinical analysis, pharmaceuticals, detergents, food production and fermentation. The use of rennet, as an exogenous enzyme, in cheese manufacture Proteinases in dairy technology - acceleration of cheese ripening, modification of functional properties and preparation of dietic products.

Biotechnology in the production of enzymes

Enzymes are biological catalysts used to facilitate and speed up metabolic reactions in living organisms. They are proteins and require a specific substrate on which to work. Their catalysing conditions are set within narrow limits, e.g. optimum temperature, pH conditions and oxygen concentration. Most enzymes are denatured at temperatures above 42°C. However, certain bacterial enzymes are tolerant to a broader temperature range.

Enzymes are essential in the metabolism of all living organisms and are widely applied as processing aids in the food and beverage industry.

In the past, enzymes were isolated primarily from plant and animal sources, and thus a relatively limited number of enzymes were available to the food processor at a high cost. Today, bacteria and fungi are exploited and used for the commercial production of a diversity of enzymes. Several strains of microorganisms have been selected or genetically modified to increase the efficiency with which they produce enzymes. In most cases, the modified genes are of microbial origin, although they may also come from different kingdoms. For example, the DNA coding for chymosin, an enzyme found in the stomach of calves, that causes milk to curdle during the production of cheese, has been successfully cloned into yeasts (*Kluyveromyces lactis*), bacteria (*Escherichia coli*) and moulds (*Aspergillus niger* var. *awamori*). Chymosin produced by these recombinant microorganisms is currently commercially produced and is widely used in cheese manufacture.

Genetic technologies have not only improved the efficiency with which enzymes can be produced, but they have increased their availability, reduced their cost and improved their quality. This has had the beneficial impact of increasing efficiency and streamlining processes which employ the use of enzymes as processing aids in the food industry.

In addition, through protein engineering, it is possible to generate novel enzymes with modified structures that confer novel desired properties, such as improved activity or thermostability or the ability to work on a new substrate or at a higher pH. Directed evolution is one of the main methods currently used for protein engineering. This technique involves creating large numbers of new enzyme variants by random genetic mutation and subsequently screening them to identify the improved variants. This process is carried out repeatedly, thus mimicking natural evolution processes.

❑❑❑

Chapter 27

Industrial Production of Aminoacids, Solvents & Vitamins

I. Amino acids

Used in food as flavour enhancers, seasonings and nutritional additives. Used to improve bread quality. Monosodium glutamate is commercially used as flavour enhancer. Glutamic acid, L-Lysine and Methionine are important amino acids.

a. Glutamic acid production

Either two stage process or one stage process is used

Substrate ----→ alpha Keto glutaric acid ----→ L glutamic acid

Organisms used

Corynebacterium glutamicum, Brevibacterium flavum, Microbacterium flavum

They require biotin for growth. Excess biotin supports abundant growth, but seriously inhibits product formation (1-5 μ gram is optimum). Glucose and starch hydrolysates are the principal carbon sources. For excretion of the produced glutamic acid through cell membrane ,biotin is used. Antibiotics can be used if the molasses are used as substrate. Temperature- 30°C, duration-40 hours.

Accumulation of glutamic acid occurs till the concentration reached 50 mg/g dry weight. If molasses is used, addition of Penicillin or detergent during the exponential phase allows excretion of glutamate.

b. L-Lysine

Essential amino acid used for supplementing cereal proteins lacking this amino acid. Auxotrophic mutant strains -*Corynebacterium glutamicum, B.flavum* are used. Production media is 1-10 % L aspartic acid. Excess amount of biotin prevents the production of glutamic acid, leads to production of L- lysine. Excess of leucine in the medium, inhibits lysine biosynthesis. Leucine auxotrophic strains produce more lysine. Addition of antibiotics inhibits the sporulation, and enhances the accumulation of lysine. Produced through diaminopimelic acid pathway.

2. Solvents

a. Acetone and Butyl Alcohol

These are two important industrial solvents produced by microbial fermentation. The process was discovered by *Chaim Weizmann* in England in the early part of this century when acetone was mainly used in the production of explosives and butanol was used in making synthetic rubber. N-butanol is used in brake fluids, in urea-formaldehyde resins and in lacquers used as protective coatings in automobiles.

Chaim Weizmann

Clostridium acetobytylicum was the first microorganism to be harnessed for industrial production of acetone from starch. *C. saccharoacetobutylicum* is capable of converting molasses to acetone and butanol. Sterile diluted molasses or cooked corn meal is used as the substrate for microbial fermentation in submerged cultures the pH being 7.2. The fermentation is anaerobic and results in the production of large amounts of hydrogen and carbon dioxide which are used as by-product. Carbon dioxide is used for the preparation of dry ice and hydrogen for fuel. Acetone, butyl alcohol and other products of fermentation are recovered by fractional distillation.

b. Glycerol

Glycerol has a number of commercial applications (including the production of explosives and propellants). Glycerol is mainly used as a solvent in food colouring agents, as a lubricant in toothpastes, candies, cake icings and in various ways in cosmetic and pharmaceutical industry.

Glycerol is produced by using yeast *Saccharomyces cerevisiae* or bacteria such as *Bacillus subtilis*.

c. Ethanol

Ethanol is produced both as a petrochemical, through the hydration of ethylene, and biologically, by fermenting sugars with yeast.

I. Chemical Ethylene hydration

Ethanol for use as industrial feedstock is most often made from petrochemical feed stocks, typically by the acid-catalyzed hydration of ethylene, represented by the chemical equation

$$C2H4 + H2O \rightarrow CH3CH2OH$$

The catalyst is most commonly phosphoric acid, adsorbed onto a porous support such as diatomaceous earth or charcoal.

II. Biological Fermentation

Ethanol can be made by a dry mill process or a wet mill process. In the dry mill process, the starch portion of the corn is fermented into sugar then distilled into alcohol.

Steps in the dry mill process

1. Milling. The feedstock passes through a hammer mill which grinds it into a fine powder called meal.
2. Liquefaction. The meal is mixed with water and alpha-amylase and then passed through cookers where the starch is liquefied. Heat is applied at this stage to enable liquefaction. Cookers with a high temperature stage (120-150 degrees Celsius) and a lower temperature holding period (95 degrees Celsius) are used. High temperatures reduce bacteria levels in the mash.
3. Saccharification. The mash from the cookers is cooled and the secondary enzyme (gluco-amylase) is added to convert the liquefied starch to fermentable sugars (dextrose).

4. *Fermentation.* Yeast (*Saccharomyces cerevisiae*) is added to the mash to ferment the sugars to ethanol and carbon dioxide. Using a continuous process, the fermenting mash is allowed to flow through several fermenters until it is fully fermented and leaves the final tank. In a batch process, the mash stays in one fermenter for about 48 hours before the distillation process is started.
5. *Distillation.* The fermented mash, now called beer, contains about 10% alcohol plus all the non-fermentable solids from the corn and yeast cells. The mash is pumped to the continuous flow, multi-column distillation system where the alcohol is removed from the solids and the water. The alcohol leaves the top of the final column at about 96% strength, and the residue mash, called stillage, is transferred from the base of the column to the co-product processing area.
6. *Dehydration.* The alcohol from the top of the column passes through a dehydration system where the remaining water will be removed. Most ethanol plants use a molecular sieve to capture the last bit of water in the ethanol. The alcohol product at this stage is called anhydrous ethanol (pure, without water) and is approximately 200 proof.
7. *Denaturing.* Ethanol that will be used for fuel must be denatured, or made unfit for human consumption, with a small amount of gasoline (2-5%). This is done at the ethanol plant.

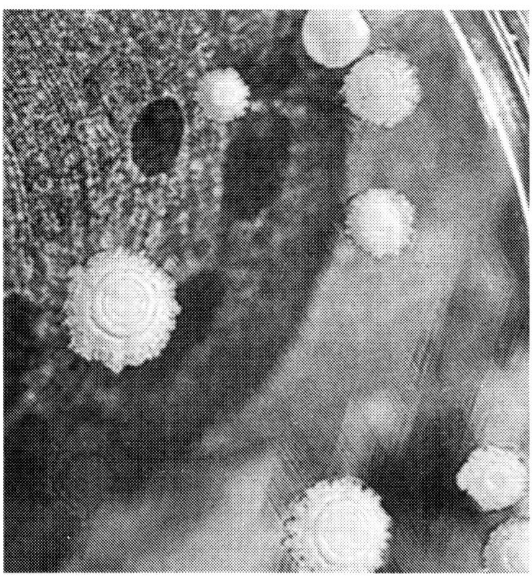

Saccharomyces cerevisiae

8. *Co-Products*. There are two main co-products created in the production of ethanol: distillers grain and carbon dioxide. Distillers grain, used wet or dry, is a highly nutritious livestock feed. Carbon dioxide is given off in great quantities during fermentation and many ethanol plants collect, compress, and sell it for use in other industries.

Most of the ethanol (ethyl alcohol) produced from microbial fermentation is used up in the preparation of beverages. Production of ethanol from sugary substrates is expensive compared to naturally available petroleum. The microbial conversion of sugar into alcohol is limited by the toxicity of alcohol which cannot accumulate in the fermenter beyond a limit. The recovery process which involves distillation requires energy input.

Ethanol production by Bacteria

Recently, Bacteria which can ferment sugar faster than yeast have been found, e.g., *Zymomonas mobillis* and the thermophillic *Thermobacter ethanolicus.* As sugary substrates are expensive and are used for food, it may be possible to use cellulosic materials and photosynthetic microorganisms. Conversion of cellulose to ethanol involves two steps, first the conversion of cellulose to sugar and second, sugar to alcohol. Species of *Clostridium* have been found of useful in the former and for the second step *Zymomonas mobillis* or *Thermobacter ethanolicus* can be employed. The anaerobic bacterium *Clostridium ljungdahlii,* discovered in commercial chicken wastes, can produce

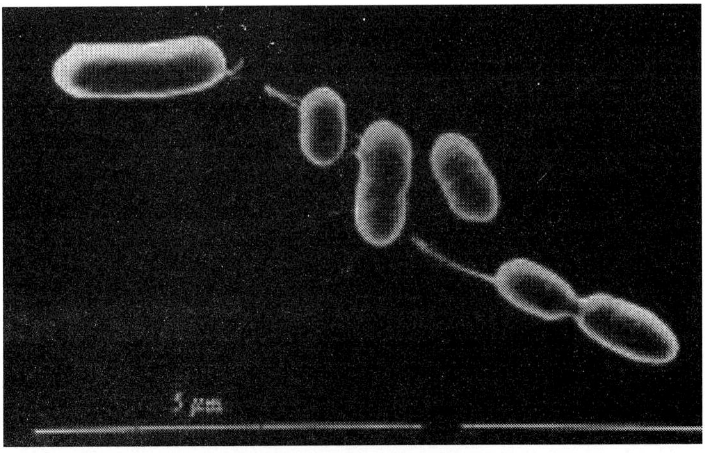

Zymomonas mobillis

ethanol from single-carbon sources including synthesis gas, a mixture of carbon monoxide and hydrogen that can be generated from the partial combustion of either fossil fuels or biomass. If these techniques are developed for industrial production coupled with genetic engineering techniques, it may be possible to produce cheap ethanol to meet the organic solvents and automotive fuel requirements of the world.

The advantages of *Z. mobilis* over *S. cerevisiae* with respect to producing ethanol,

- Higher sugar uptake and ethanol yield
- Lower biomass production
- Higher ethanol tolerance
- Does not require controlled addition of oxygen during the fermentation
- Amenability to genetic manipulations

However, it has a severe limitation compared to yeast:

(i) Its inability to convert complex carbohydrate polymers like cellulose, hemicellulose, and starch to ethanol. Its utilizable substrate range is restricted to glucose, fructose, and sucrose. Using biotechnological methods, scientists are currently trying to overcome this. A variant of *Z. mobilis* that is able to use certain pentoses as a carbon source has been developed.

(ii) Its resulting in byproducts such as sorbitol, acetoin, glycerol, and acetic acid

(iii) Formation of extracellular levan polymer.

Research works going on in eliminating these defects.

3. Vitamins

Several vitamins are produced during the normal metabolism of microorganism. All the m.o will not accumulate the vitamins in the growth medium and it is utilized for their growth. Some excrete vitamins in excesss. Riboflavin (B_2), B_{12} and vit C have found commercial production through microbial fermentations.

a. Vitamin B_{12} (cobamide)

Group of closely related cobamide compounds

Cobamides all contain a cobalt porphyrin nucleus to which is attached ribose and phosphate

Vitamin B_{12} as recovered from liver preparations is know as cyanocobalamin.

Fermentation process

- Produced by submerged culture technique
- By product of Streptomycin antibiotic fermentation ,acetone butanol fermentation
- Also produced by direct fermentations using *Streptomyces olivaceous*
- Inoculum medium need not be prepared with precursor
- Crude medium with glucose as C source, Cobalt chloride as precursor (2-10 ppm) and crude N source is used for production
- Aeration, agitation is required. Antifoam agents are needed
- 3-5 days duration
- Lysis of the mycelium begins ,pH increases, and release of bound vitamins occurs

Microbes: *Streptomyces griseus, S.olivaceous, Bacillus megaterium, B.coagulans, Pseusomonas denitrificans, Propionibacterium freudenreichii*

Recovery: During fermentation most of the product is associated with mycelium and at the end it is released to the broth

- Acidification and heating releases the product into the broth
- Broth is filtered and treated with cyanide to bring about the conversion of cobalamine to cyanocobalamine
- It is then adsorbed in packed columns
- Then it is eluted using solvents
- Precipitated as copper or zinc cyanide cobalamine

b. Riboflavin (B_2)

Riboflavin is a vitamin essential for the growth and reproduction of both human and animals. It is used for feed additive for the nutrition of various domestic animals. Commercial production is by microbial fermentation. Also produced as by product of the acetone-butanol fermentation.

Microorganisms

- By-product of the acetone-butanol fermentation - *Clostridium butylicum, Clostridium acetobutylicum.*
- Produced by direct fermentation - *Eremothecium ashbyii* and *Ashbya gossypii.*
- *Candida species* like *Candida guilliermondia and Candida flareri* produced good yield- not commercially employed.

Fermentation process

- *Eremothecium ashbyii and Ashbya gossypii* are used
- Varies in production stability
- Media-semi purified sugars (Glucose) with crude organic nutrients
- Corn oil can also be used instead of glucose
- pH 6-7.5, 26-28°C, aerated submerged fermentation, 4-5 days
- Autolysis of the cells and release of bound riboflavin
- Rapid cell growth, occurs and conversion of glucose to pyruvic acid
- Glucose depletion leads to growth cease and sporulation begins decrease in pyruvic acid
- Accumulation of ammonia, rapid synthesis of cell bound riboflavin
- Cell autolysis and release of bound riboflavins

Recovery

- Heating of the broth at 120°C for 1 hr
- Filtration
- Precipitation by adding dithionate
- Purification
- Crystallization

Sucking of Agave plant sap for Pulque

Other fermented sources of vitamins

- *Sorghum beer* in Southern Africa contains relatively high levels of riboflavin and nicotinic acid, which are important for people consuming a high maize diet. *Pellagra* (a vitamin deficiency disease associated with high maize diets) is unusual in communities in which sorghum beer is consumed. Even children benefit from consuming the dregs which contain relatively little alcohol but are rich in vitamins.
- *Palm wine* in West Africa is rich in vitamin B12. This is very important for people with low meat intake, and who subsist primarily on a vegetarian diet.
- *Pulque* (a fermented Agave plant sap) is an important source of vitamins for the economically deprived people in Mexico. The fermentation process involved in Pulque production increases its vitamin content. For instance the vitamin content (milligrams of vitamins per 100g of product) of pulque increases from 5 to 29 for thiamine, 54 to 515 for niacin and 18 to 33 for riboflavin during fermentation.
- *Idli* (a lactic acid bacteria fermented product consumed in India) is high in thiamine and riboflavin.

Vitamins obtained from microbial fermentations

Vitamin	**Organism**	**Medium**	**Fermentation Conditions**	**Yield**
Riboflavin	*Ashbya gossypii*	Glucose, collagen, Soya oil, Glycine	6 days at 36°C; Aerobic	4.25 g/L
L-Sorbose (for Vitamin C)	*Gluconobacter oxidans* sub sp. *suboxidans*	30% Cornsteep, D. Sorbitol.	45 hrs at 30°C Aerobic	70% (based on substrate used)
5-Ketogluconic acid (for Vitamin C)	*Gluconobacter oxidans* sub sp. *suboxidans*	Glucose, CaCO3, cornsteep.	33 hrs. at 30°C; Aerobic	100% (based on substrate used)
Vitamin B_{12}	*Propionibacterium shermanii*	Glucose, Cornsteep, ammonia, cobalt, pH 7.0	3 days at 30°C anaerobic and 4 days aerobic	23 mg/L

❒❒❒

Chapter 28

Fermented Foods — Alcoholic Beverages

The use of natural biological processes to obtain useable products is certainly not new. Since recorded history, microbes have been involved in the preparation and processing of items in the daily diet of humans. Occasionally, however, microbial infections of food materials made it appear more appetizing and the taste enhanced. Ultimately, microbial infections of these foods were exploited, so the fermented foods and beverages now form a large and important sector of the food industry. Today the main groups of microbes involved in the industry include the yeasts, molds, and bacteria. Important fermented foods fall under 4 major groups

- Alcoholic Beverages – Beer, wine, distilled spirits
- Dairy Products – Cheese, Yoghurt, Curd, butter, ghee
- Cereal products – Bread, Idli
- Other Fermented Foods – Pickles, Sauerkraut
- Microbes as food – SCP (Single Cell Protein)

I. Alcoholic Beverages

A. Brewing Beer

Beer is a fermentation of malted grains (especially barley) and hops by yeast. The starch in barley is broken down into glucose and then fermented to ethanol by yeast. The finish product is aged and then packaged for distribution and consumption. It involves six stage process, beginning with the formation of malt from barley.

1. *Malt* - Barley is first soaked in water for 5 to 7 days. At this time the grain germinates and produces amylases (enzymes that degrade starch to glucose) and proteases (enzymes that break down proteins). These enzymes are essential to the brewing process. Amylase provides sugar for the yeast fermentation and the proteases solubilize compounds in the grain and hops important for the quality of the beer. The germinated malt is then dried and crushed.

2. *Mash* - Mashing solubilizes the starch and other flavors in the grain and extracts flavors and preservatives for the beer. The prepared malt is suspended in water mixed with boiled malt adjuncts (other grains, carbohydrates and sugars that provide a source of starch to be converted to sugar). This mash is then incubated at 65-70°C for a short time to allow the amylase to break down the starch to glucose. The temperature is raised to 75°C to inactivate the enzymes and then allowed to settle. Insoluble matter sinks to the bottom and serves as a filter as the liquid (now called wort) is taken from the container.

3. *Boiling with Hops* – Hops are nothing but the scales of female flowers of Hop vine. Hops and wort are combined and boiled for 2.5 hours. The liquid is removed and ready for fermentation. Boiling with hops serves several purposes -

- Concentration
- Sterilization, killing many microbes that might spoil the beer
- Further inactivation of enzymes in the mash.
- Solubilization of important compounds in the hops and mash. Some of these add to the flavor of the beer while others, especially from the hops, have antiseptic qualities and help to preserve the beer.

Malting

Mashing

Addition of Hops

Fermentation

Aging

Stages of Beer Fermentation

4. *Fermentation* : Fermentation begins by adding the brewers yeast *Saccharomyces carlsbergensis* to the wort. The starter culture is usually obtained from a previous batch of beer and is added at a very high concentration (500 grams per 120 liters). Fermentation is at a low temperature between 3.3 and 14°C for 8 to 14 days. At this time the glucose in wort is converted to ethanol and CO_2. Other compounds in the wort are also fermented to add to the characteristic flavor of beer.

5. *Aging* : The fermented wort (green beer) is aged at 0°C for a period of weeks or months depending on the brewer. At this time the yeast settle to the bottom of the vessel, bitter flavors are mellowed and other compounds are formed that enhance flavor.

6. *Finishing* : The beer is now prepared for packaging. This can involve filtering, pasteurization, carbonation to 0.45 to 0.52% CO_2, and clarification. All of these processes depend upon the beer being made and each brewery will specialize the fermentation, aging and finishing of their beer. This is often the inspiration for various advertising done by the brewery.

   The beer is then put in containers and distributed to customers. Beer normally has a shelf life of about 6 months and after that starts to take on undesirable flavors.

B. Wine

Wine is produced from the fermentation of fruit juice, usually from grapes

1. Preparation of MUST

The grapes are crushed to form a "**must**"

For white wines, white grapes are usually used, and the skins are removed from the must ("**pressing**") before fermentation. For red wines, red or black grapes are used, and the skin is allowed to remain during fermentation. For rosé wines, red grapes are used and the juice is allowed to remain in contact with the skins just long enough for a rose or pink color to develop

2. Primary fermentation

Natural yeasts on the skins of the grapes may be used, but in commercial production cultured yeast is often used to give more predictable results. The **yeast (*Sacchromyces ellipsoides*)** will turn the sugar in the juice into alcohol and carbon dioxide. Fermentation is carried out in stainless steel vessels. The temperature should be maintained below 30°C.

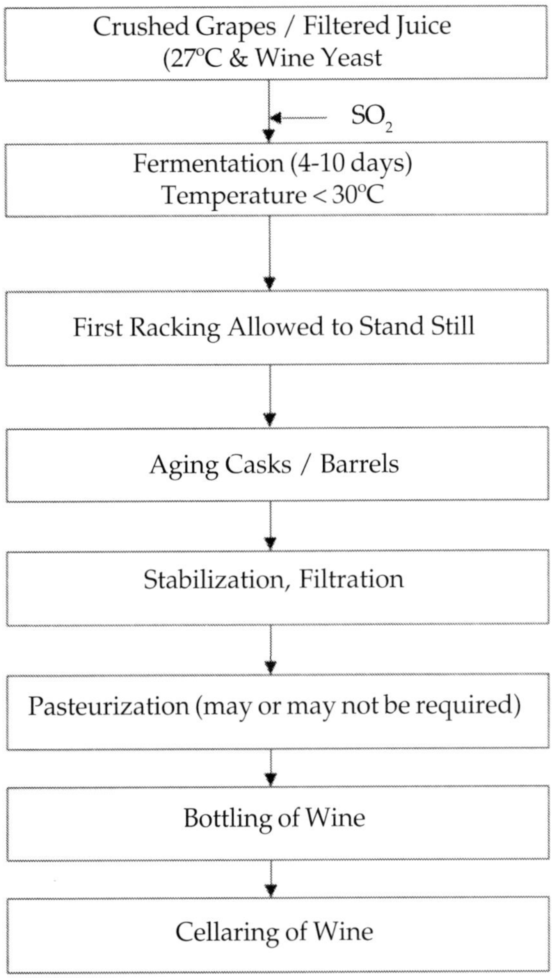

After primary fermentation, the must is pressed (red wines) and transferred to different containers for secondary fermentation.

3. Secondary fermentation

The winemaker may choose to allow a white wine to undergo a second fermentation which occurs due to malic acid in the grape juice. The vessel is kept airtight to prevent oxidation. When malic acid is allowed to break down into carbon dioxide and lactic acid, it is known as *"malo-lactic fermentation,"* which imparts additional flavor to the wine. A "buttery" flavor in some whites is due to this process. This process is used for sparkling wines. Takes 3 – 6 months. It is done in either stainless steel vessels or in oaken barrels.

4. SO_2 addition

A small amount of SO_2 is produced naturally as a by product of fermentation, but most of the SO_2 has been added by the winemaker @ 30 mg/litre. During white wine production, it is added at almost every stage of the process, and is more or less required after malolactic fermentation is complete.

Purpose of SO_2 addition

1. *As Anti microbial Agent* : The most important mechanism of action for Sulfur Dioxide is as an anti-microbial agent. It regulates the growth of harmful yeast and bacterial growth in the wine.
2. *As anti-oxidant* : This guards against browning and protects the fruit-like qualities of the wine. SO_2 can bind with a molecule called acetaldehyde. It is produced when a wine undergoes oxidation. When SO_2 reacts with acetaldehyde, they bond together, producing a harmless, odorless molecule

5. First Racking

After fermentation completed naturally or stopped by addition of distilled spirit, first racking is carried out. This involves the wine to stand still until most yeast cells and fine suspended material settle out. The wine is then filtered without disturbing the sediment or the yeast.

6. Aging

Aging is done to add additional *clarification* and more complex *flavors*. Flavor can come from wood. The wine may be barrel aged for several months to several years. No air is allowed to enter the barrels during this period. During the winery aging, the smaller containers may develop differences. So the winemaker will probably "blend" wine from different barrels, to achieve a uniform result.

7. Stabilization and filtration

Stabilization is carried out by chilling the wine *to remove* traces of *tartaric acid*. These tartarates present in the grape juice tend to crystallize in wine. These crystals can be removed by filtration

8. Pasteurization

If the wine has an alcohol content **less than 14%** it may be heat pasteurized or cold pasteurized through microporous filters just before bottling.

9. Bottling

Producers often use different shaped bottles to denote different types of wine. *Colored bottles* help to reduce damage by light. Light causes oxidation and breakdown of the wine into chemicals, such as mercaptan, which are undesirable. The life of wine in the bottle is longer if the pH is less than 3.4 to 3.45.

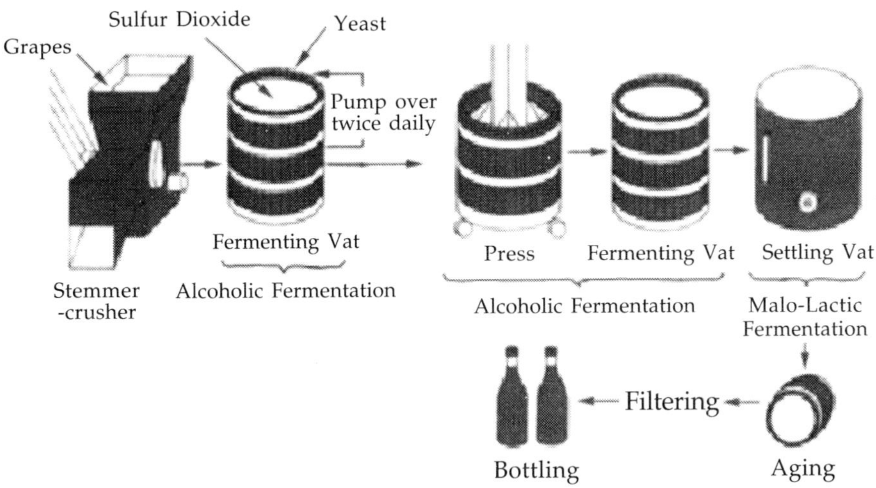

Types of wine

a. Dry vs. sweet

- In dry wine, all the sugar has been converted to ethanol.
- Sweet wine still has a lot of the sugars from the fruit juice. It is prepared by doing only short fermentations or by using yeasts with low alcohol tolerances so that not all the sugar is consumed.

b. Champagne vs. sparkling wines

- Production of champagne involves a final in-bottle fermentation, that generates the carbon dioxide in the final solution. The yeasts that settle in the long neck of the bottle, which is inverted during this process, are frozen and removed. A small amount of sugar is added, and the bottle is tightly corked.

- In sparkling wines, carbon dioxide is injected. Consequently, sparkling wines lack the flavors from having the yeasts do the final fermentation, as in champagne.

c. White vs. red

- White wines use both red and white grapes, but the skins are removed after a short time of fermentation. White wines are usually not aged — generally, they are sold after five months.
- Red wines are made from red grapes only, and the skin is kept in the fermentation mixture for longer times before being removed. Red wines undergo several years of aging in barrels before being bottled and sold.

d. Young vs. mature

- In mature wines, the malate in grape juice is converted to ethanol by *Lactobacillus*. This makes mature wines less acidic and more flavorful compared to young wines.
- In addition, many other chemical reactions occur during aging, most of which are not yet understood.

e. Bad wine

If the cork is allowed to dry, air and microorganisms may enter the bottle causing the following reaction:

$$CH_3CH_2OH + 2\ NAD^+ \longrightarrow 2\ CH_3COO^- + 2\ NADH$$

Generation of acetic acid (CH_3COO^-) from ethanol amounts to the production of vinegar. Thus the wine will be *soured*. This incomplete fermentation of ethanol to acetic acid, which requires *aerobic* conditions, is catalyzed predominantly by bacteria of the genera *Acetobacter* and *Gluconobacter*.

C. Distilled spirits

Produced by the fermentation of grain mash (similar to beer), followed by distillation to increase the alcohol content. Different types of grain are used to produce different types of distilled spirits.

❑❑❑

Chapter 29

Fermented Foods-Dairy Products

Many products are made through microbial fermentation of milk, including buttermilk, yogurt, cheese, kefir, kumiss etc. Fermentation is primarily carried out by lactic acid bacteria. The lactic acid pathway and the accumulation of lactic acid from the metabolism of milk sugar, lactose are common to the production of fermented dairy products. The differences in the flavour and aroma of the various dairy products are due to additional fermentation products, which may be present in very low concentrations.

a. Yogurt

Yogurt is nothing but curd. It is produced by the controlled fermentation of milk by two species of bacteria (*Lactobacillus* sp. and *Streptococcus* sp.). The sugar in milk (called lactose) is fermented to acid (lactic acid) and this causes the characteristic curd to form. The acid also restricts the growth of food poisoning bacteria and some spoilage bacteria. So, whereas milk is a potential source of food poisoning and only has a shelf life of a few days, yoghurt is safer and can be kept for up to ten days, under proper storage conditions.

Yoghurt can be easily produced at the small-scale. The procedure is as follows:

- Production of yogurt starts by conditioning the milk. The water content of milk is first lowered to 25% by vacuum evaporation and 5% milk solids are added.
- As a final conditioning step, the milk is heated to 86 to 93°C for 30-60 minutes. This causes some breakdown of proteins and other molecules and kills contaminating microbes that may compete with the starter culture.
- After cooling to 45°C a 1:1 mixture of *Streptococcus thermophilus* and *Lactobacillus bulgaricus* is added.
- Fermentation is at 45°C until the desired degree of acidity is reached. This usually occurs in 3-5 hours.
- The finished product may have other ingredients added (such as mold inhibitors or dye). Yogurt is stored at 0-4°C until consumed to prevent spoilage.

There is some evidence that consumption of products containing active cultures of lactic acid bacteria can be beneficial. It is reasonable to assume that ingesting lactic acid bacteria may help deter other more severe pathogens such as *E. coli* or *Salmonella typhimurium.*

b. Cheese

Cheese is also a milk fermentation, but its production is more complex. Different bacteria come into play and production periods are much longer than yogurt. Despite there being 20 classes and hundreds of varieties of cheeses the initial manufacturing process is surprisingly similar.

In general cheese production involves the following steps.

Steps involved in the Preparation of cheese

1. Standardize Milk

Milk is often standardized before cheese making to optimize the protein to fat ratio to make a good quality cheese with a high yield.

2. Pasteurize/Heat Treat Milk

Depending on the desired cheese, the milk may be pasteurized or mildly heat-treated to reduce the number of spoilage organisms and improve the environment for the starter cultures to grow. Some varieties of milk are made from raw milk so they are not pasteurized or heat-treated. Raw milk cheeses must be aged for at least 60 days to reduce the possibility of exposure to disease causing microorganisms (pathogens) that may be present in the milk.

3. Cool Milk

Milk is cooled after pasteurization or heat treatment to 90°F (32°C) to bring it to the temperature needed for the starter bacteria to grow. If raw milk is used the milk must be heated to 90°F (32°C).

4. Inoculate with Starter & Non-Starter Bacteria and Ripen

The starter cultures are added to the milk and held at 90°F (32°C) for 30 minutes to ripen. Generally *Lactococcus lactis, Lactococcus cremoris* and *Lactococcus cremoris* subsp. *diacetylactis* are used. The former two bacteria are responsible for lactic acid fermentation and the last bacteria inculcates & favour to the product. The ripening step allows the bacteria to grow and begin fermentation, which lowers the pH and develops the flavor of the cheese.

5. Add Rennet and Form Curd

The rennet is the enzyme that acts on the milk proteins to form the curd. After the rennet is added, the curd is not disturbed for approximately 30 minutes so a firm coagulum forms.

6. Cut Curd and Heat

The curd is allowed to ferment until it reaches pH 6.4. The curd is then cut with cheese knives into small pieces and heated to 100°F (38°C). The heating step helps to separate the whey from the curd.

7. Drain whey

The whey is drained from the vat and the curd forms a mat.

8. Texture curd

The curd mats are cut into sections and piled on top of each other and flipped periodically. This step is called *cheddaring*. Cheddaring helps to expel more whey, allows the fermentation to continue until a pH of 5.1 to 5.5 is reached, and allows the mats to "knit" together and form a tighter matted structure. The curd mats are then milled (cut) into smaller pieces.

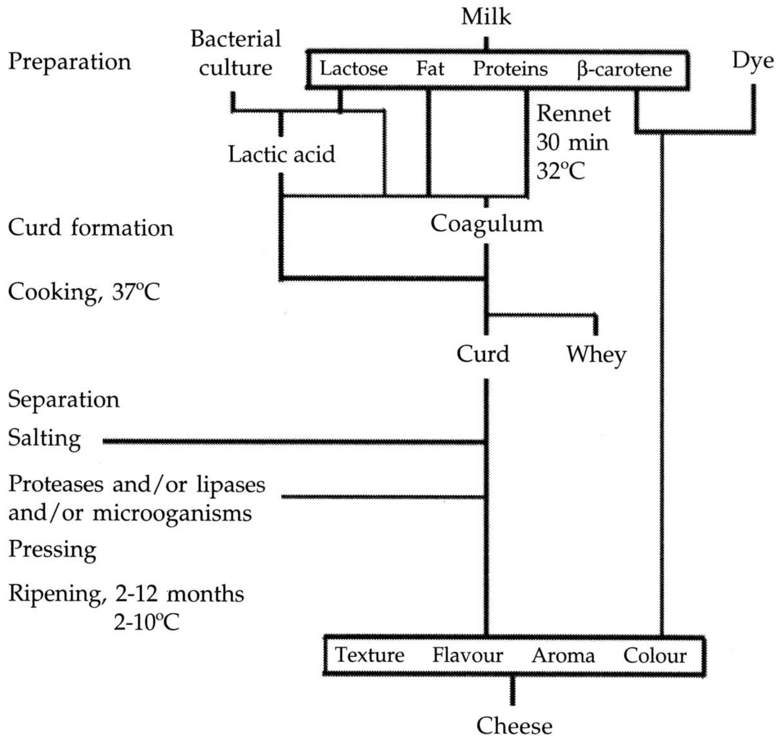

9. Dry Salt or Brine

For cheddar cheese, the smaller, milled curd pieces are put back in the vat and salted by sprinkling dry salt on the curd and mixing in the salt. In some cheese varieties, such as mozzarella, the curd is formed into loaves and then the loaves are placed in a brine (salt water solution).

10. Form Cheese into Blocks

The salted curd pieces are placed in cheese hoops and pressed into blocks to form the cheese.

11. Store and Age

The cheese is stored in coolers until the desired age is reached. Depending on the variety, cheese can be aged from several months to several years.

12. Package

Cheese may be cut and packaged into blocks or it may be waxed.

Types of cheeses

Cheese cannot be categorized on a single categorization concept. Therefore, various systems are used for the categorization of cheese. Some factors taken into consideration while classifying cheese are the length of aging, the methods of making it, the curd and the various processes relating to the curd, the kind of milk, the fat content and the texture.

Based on their firmness they are classified as soft, semisoft and hard.

A. *Soft cheese :*

Eg.

i. Brie cheese - mild tangy and fruity flavoured
ii. Cream cheese - made from raw unpastuerised milk in small batches by hand
iii. Mascarpone cheese - Italian rich cheese used in deserts
iv. Camembert cheese - *Penicillium camemberti*

B. *Semi Soft cheese:*

Eg.

i. Roquefort cheese, - *Penicillium roqueforti*
ii. Blue cheese-white cheese over which a blue mold is dusted. During aging the mold punctures into the cheese. It has sharp and tangy flavour.

C. Hard cheese:

Eg.

i. Cheddar cheese
ii. Swiss cheese - *Propionibacterium shermani*
iii. Parmesan cheese
iv. Colby cheese

c. Butter

Butter is a dairy product made by churning fresh or fermented cream or milk. Butter consists of butterfat surrounding minuscule droplets consisting mostly of water and milk proteins. The most common form of butter is made from cows' milk, but can also be made from the milk of other mammals, including sheep, goats, buffalo, and yaks.

Steps in Butter Preparation

1. Pasteurized cream is fermented by adding *Lactococcus* and *Leuconostoc* bacteria. During fermentation, the cream naturally sours as bacteria convert milk sugars into lactic acid. The fermentation process produces additional aroma compounds, including diacetyl, which makes a good flavor and taste.
2. This pasteurized cream is heated to a relatively high temperature above 80°C (180°F).
3. The cream is cooled to about 5°C (40°F) and allowed to remain at that temperature for at least eight hours; under these conditions about half the butterfat in the cream crystallizes.
4. The fermented cream then churned. Churning produces small butter grains floating in the water-based portion of the cream. This watery liquid is called buttermilk
5. The buttermilk is drained off; sometimes more buttermilk is removed by rinsing the grains with water.
6. Then the grains are "worked": pressed and kneaded together. When prepared manually, this is done using wooden boards called scotch hands. This consolidates the butter into a solid mass and breaks up embedded pockets of buttermilk or water into tiny droplets.

Commercial butter is about 80% butterfat and 15% water; traditionally-made butter may have as little as 65% fat and 30% water.

Types of Butter

1. *Cultured butter*- Butter made from a fermented cream is known as cultured butter. Made from pasteurized cream whose fermentation is produced by the introduction of *Lactococcus* and *Leuconostoc* bacteria. Sometimes labeled *European-style butter*
2. *Sweet cream butter* - Butter made from pasteurized fresh cream is called sweet cream butter. Sweet cream butter dominates in the United States and the United Kingdom
3. *Raw cream butter* - Butter made from fresh or cultured unpasteurized cream is called raw cream butter.
4. *Spreadable butter* - spreadable butter remain softer at colder temperatures and are therefore easier to use directly out of refrigeration.
5. *Whipped butter* - Another product designed to be more spreadable, is aerated via the incorporation of nitrogen gas—normal air is not used, because doing so would encourage oxidation and rancidity.
6. *Clarified butter* is butter with almost all of its water and milk solids removed, leaving almost-pure butterfat. Clarified butter is made by heating butter to its melting point and then allowing it to cool off; after settling, the remaining components separate by density. At the top, whey proteins form a skin which is removed, and the resulting butterfat is then poured off from the mixture of water and casein proteins that settle to the bottom.

d. Ghee

It is a type of clarified butter important in Indian cuisine. Ghee is made by simmering unsalted butter in a large pot, until its water has boiled off and its protein has settled to the bottom. The clarified butter is then spooned off, taking care not to disturb the milk solids on the bottom of the pan. Unlike butter, ghee can be stored for extended periods without refrigeration, provided it is kept in an airtight container to prevent oxidation and that it remains moisture-free.

Preparation of Ghee

Milk is curdled. The curd is then manually churned until it precipitates butter and leaves behind some whey. The butter is then heated on a low flame until a layer of white froth covers the surface. This state indicates the end of process and the liquid obtained on filtering the suspension is pure ghee.

e. Acidophilus milk

It is produced by inoculating *Lactobacillus acidophilus* into sterile skimmed milk. The inoculum of 1-2% is added followed by holding at 37°C until a smooth curd develops.

f. Bulgarian milk

This has the origin of Bulgaria. It is also produced in the same way as that of Acidophilus milk but the starter culture is *Lactobacillus bulgaricus*.

g. Kefir

Kefir is produced by using kefir grains as starter culture. Kefir grains are hard grain like bacterial clots containing *Streptococcus lactis, Lactobacillus bulgaricus* and lactose fermenting yeast called *Saccharomyces delbreukii* held together by layers of coagulated proteins. Whole milk from cow, goat or sheep is inoculated with kefir grains and fermentation is carried out in leather bags made of goat skin at a temperature of 20 °C. The bacteria produce acid while yeast produce alcohol and concentration of acid alcohol may be 0.5 to 1.0 % Kefir grains can be dried and stored for long time.

h. Kumiss

The process is similar to kefir but made of mare's milk. Culture do not form grains and alcohol content may reach 2%.

❑❑❑

Chapter 30

Fermented Foods-Cereal Products and Others

1. Fermented Cereal products

A. Bread

Bread is a simple fermentation of sugar to CO_2 and alcohol. The baker first combines flour, sugar, milk and other ingredients with a microorganism, usually a bread yeast such as *Saccharomyces cerevisiae*, but not always. The ingredients are mixed and then allowed to incubate at 27°C for a few hours. During this time the yeast convert the sugar present to ethanol and CO_2. Most incubations are for less than 4 hours not leaving enough time for the yeast to increase in number. The CO_2 produced causes the bread to rise (leaven) and become porous. The success of leavening is dependent upon the rate of gas production. This can be increased by adding more yeast, more sugar, or dough conditioners (various salts that the yeast need). Adding too much of anything can either kill the yeast or cause the bread to rise too quickly. The temperature of incubation is another critical consideration. *Saccharomyces* grows best at 26 to 28°C and deviations from that temperature will usually result in slow or complete lack of leavening.

B. Idli

It is a South Indian dish. Prepared by mixing parboiled rice and blackgram in the ratio of 3:1 or 4:1, soaking for 3-10 hours and the ground into a paste. The paste is allowed to undergo fermentation overnight (10-12 hours) at temperature of 25-30°C. the batter increases in volume due to fermentation by naturally occurring bacteria *Leuconostoc mesenteroides, Pediococcus cerevisiae* and *Streptococcus faecalis.* During fermentation lactic acid and acetic acid gets accumulated and the batter becomes soured. Then this batter is steamed to get idli.

2. Other Fermented foods

i. Fermented Vegetables

Fermentation has been utilized for centuries to preserve vegetables because fresh vegetables are highly perishable. As early as the third century BC, the Chinese described the preservation of vegetables by fermentation. Vegetable preservation is achieved by the addition of salt and or by the fermentative conversion of plant sugars to organic acids by LAB (Lactic Acid Bacteria). Preservation by fermentation is simple and energy efficient. Fermented vegetables can often be stored for a year or more without serious loss of flavor or texture.

A. Sauerkraut

Shredded cabbage or other suitable vegetables are placed in a jar and salt is added. Mechanical pressure is applied to the cabbage to expel the juice, which contains fermentable sugars and other nutrients suitable for microbial activity. The first micro-organisms to start acting are the gas-producing cocci *(Leuconostoc mesenteroides).* These microbes produce acids. When the acidity reaches 0.25 to 0.3% (calculated as lactic acid), these bacteria slow down and begin to die off, although their enzymes continue to function. The activity initiated by the *L. mesenteroides* is continued by the lactobacilli *(L. plantarum and L. cucumeris)* until an acidity level of 1.5 to 2% is attained. The high salt concentration and low temperature inhibit these bacteria to some extent. Finally, *L. pentoaceticus* continues the fermentation, bringing the acidity to 2 to 2.5% thus completing the fermentation.

The end products of a normal kraut fermentation are lactic acid along with smaller amounts of acetic and propionic acids, a mixture of gases of which carbon dioxide is the principal gas, small amounts of alcohol and a mixture of aromatic esters. The acids, in combination with alcohol form esters, which contribute to the characteristic flavour of sauerkraut. The acidity helps to control the growth of spoilage and putrefactive organisms and contributes to the extended shelf life of the product. Changes in the sequence of desirable bacteria, or indeed the presence of undesirable bacteria, alter the taste and quality of the product.

B. Pickles

These include the fruit and vegetable pickles produced by lactic acid fermentation and the products of alkaline bacterial fermentations. Pickles can be made by storing prepared vegetables in a weak brine solution, by dry salting or allowing the vegetables to ferment without salt.

a. Dry Salted Pickles

Dry salting is used for pickling many vegetables and fruits including limes, lemons and cucumbers. For dry salt pickling any variety of common salt is suitable as long as it is pure.

(i) Dry salted lime pickle

Limes are placed in a layer, approximately 2.5 cm deep, into the fermenting container (a barrel or keg). One kilogram of salt is added for every four kilograms of limes. The salt is sprinkled over the vegetables. Another layer of vegetables is added and more salt added. This is repeated until the container is three quarters full. A cloth is placed above the vegetables and a weight added to compress the vegetables and assist in the formation of a brine. The formation of a brine takes about 24 hours. The container is then placed in the sun for a week.

As soon as the brine is formed, fermentation starts. As fermentation starts bubbles of carbon dioxide appear. Fermentation takes between one and four weeks depending on the ambient temperature. Fermentation is complete when no more bubbles appear.

(ii) Pickled cucumbers

One kg of salt is added to every 20 kg of small cucumbers and 15 kg of large cucumbers. The brine should be formed within 24 hours by osmosis. If the brine formed by osmosis does not cover the cucumbers 40° Salometer brine is added to the desired level. A day or two after the tank is filled and closed. The brine should be stirred in order to help equalize the concentration of salt throughout the mass. Fermentation is complete when no more bubbles appear. During fermentation the brine becomes cloudy for the first few days due to the growth of bacteria. Later if the brine is not covered, a filmy yeast growth will often occur on the surface.

(iii) Pak-Gard-Dong (Pickled leafy vegetable)

Pak-Gard-Dong is fermented mustard leaf .The mustard leaves are washed, wilted in the sun, mixed with salt, packed into containers for 12 hours. The water is then drained and a 3% sugar solution added. They are again allowed to ferment for three to five days at room temperature. Micro-organisms associated with the fermentation include *Lactobacillus brevis, Pediococcus cerevisiae* and *Lactobacillus plantarum.*

(iv) Tempoyak (pickled durian)

Tempoyak is the fermented pulp of a durian fruit from Malaysia. It has the distinctive durian smell and a creamy yellow colour. Fermentation takes about seven days.

(v) Pickled beetroots

In Russia beetroot is pickled by cleaning, slicing and placing in a container with salt. Due to the high sucrose level, dextrans are produced giving the product a slimy texture.

(vi) Green Mango Pickle

The sliced mangoes are soaked in brine solution. Sodium metabisulphite (1000 ppm) and 1% calcium chloride are added. The containers are stored until the mangoes are pickled.

(vii) Kimchi (pickled cabbage)

The prepared cabbage is placed in a salt solution (8-15%) for two to seven hours. *Lactobacillus plantarum, L. Brevis, Streptococcus faecalis, Leuconostoc mesenteroides* and *Pediococcus pentosaceus,* are the microbes involved.

b. Non Salted Pickles

(i) Gundruk (pickled leafy vegetable)

Mustard, radish and cauliflower leaves are allowed to wilt for one or two days and then shredded and added with warm water, kept for 5-7 days.

(ii) Kocho (pickled false banana)

Root crops, breadfruit and bananas are peeled before placing in the pit, which is lined with green banana leaves and left to ferment for three to six weeks, after which time it becomes soft, has a strong odour and a paste-like consistency.

(iii) Sinki (pickled radish)

Sinki is a sour pickle prepared from radish tap roots. Preparation is as that of Gundruk but kept for 12 days. *Sinki* fermentation is initiated by *Lactobacillus fermentum* and *L. brevis*, followed by *L. plantarum*. During fermentation the pH drops from 6.7 to 3.3. After fermentation, the radish substrate is sun-dried to a moisture level of about 21%.

(iv) Sunki

*Sunk*i is a non-salted and fermented vegetable product prepared from the leaves of turnip in Japan. The turnip is boiled, inoculated with "*Zumi*" (a wild small apple) dried *Sunki* from the previous year and allowed to ferment for one to two months. *Sunki* is produced under low temperature (in winter season). Micro-organisms involved include *Lactobacillus plantarum, L. Brevis, Bacillus coagulans and Pediococcus pentosaceus*.

(v) Kanji

After thorough washing the carrots are finely grated. Each kilogram of grated carrot is mixed with 7 litres of water, 200g of salt, 40g of crushed mustard seed and 8g of hot chilli powder and placed in a glazed earthenware vessel, which is almost entirely sealed, leaving then allowed to ferment for seven to ten days.

❑❑❑

Chapter 31

Microbes as Food — Single Cell Protein

There are some microbial products wherein the microbe itself serves as an important part in human nutrition and health. One such familiar product is Single Cell Protein (SCP).

Single Cell Protein

The term single cell protein (SCP) refers to dead, dry cells of microorganisms such as yeast, bacteria, fungi and algae which are rich source of protein. After grown on different carbon sources, the whole organism is harvested and consumed, rather than using the products of their fermentations or other processing. The name "single cell protein" was used for the first time, twenty years ago by the M.I.T. professor Carol Wilson to give a better image than "microbial protein".

Advantages of SCP over conventional protein sources

Usage of microorganisms as protein source has advantages over other protein sources. They are,

1. Small doubling time of microbial cells when compared to other protein sources as shown in table

Organism	Mass Doubling
Bacteria and yeast	10-120 min
Mold and algae	2-6 h
Grass ad some plants	1-2 wk
Chickens	2-4 wk
Pigs	4-6 wk
Cattle	1-2 mo
People	0.2-0.5 yr

Due to this property, the productivity of protein production from micro-organisms is greater than that of traditional proteins

2. It is independent of land and climate
3. It works on a continuous basis
4. It can be genetically controlled
5. It causes less pollution.

Disadvantages

There are five factors that impair the usefulness of SCP:

1. Non digestible cell wall (mainly algae)
2. High nucleic acid content
3. Unacceptable coloration (mainly with algae)
4. Disagreeable flavour (part in algae and yeasts)
5. Cells should be killed before consumption.

Thus SCP is treated with various methods in order to kill the cells, improve the digestibility and to reduce the nucleic acid content.

The Problem of Nucleic Acids

The problem which occurs from the consumption of proteins with high concentration of nucleic acids (18-25 g/100 g protein dry weight) is the high level of uric acid in the blood, sometimes resulting in the disease *gout*. Uric acid is a product of purine metabolism. Most mammals, reptiles and molluscs possess the enzyme uricase, and the end product of purine metabolism is allantoin. Man, birds and some reptiles lack the enzyme uricase and the end product of purine degradation is uric acid. The removal or reduction of nucleic acid content of various SCP's is achieved with one of the following treatments.

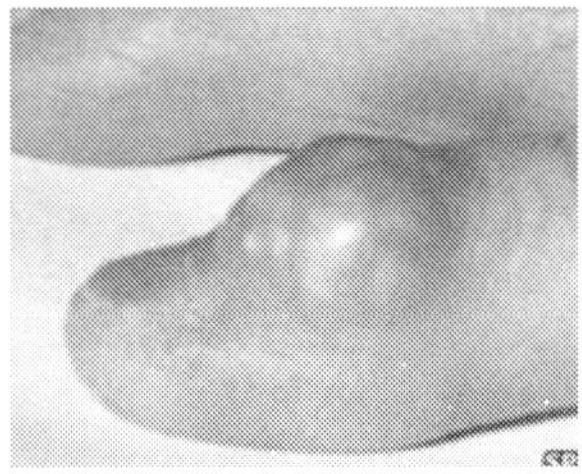

Gout

a. Treatment of cells with 10% NaCl;

b. Chemical treatment with NaOH;

c. Thermal shock.

These methods aim to reduce the RNA content from about 7% to 1% which is considered within acceptable levels.

Nutritional Value of SCP

For the assessment of the nutritional value of SCP, factors such as nutrient composition, amino acid profile, vitamin and nucleic acid content as well as palatability, allergies and gastrointestinal effects should be taken into consideration. Also long term feeding trials should be undertaken for toxicological effects and carcinogenesis.

The following shows the average cell composition of the major groups of micro-organisms (% dry weight).

	Fungi	Algae	Yeasts	Bacteria
Protein	30-45	40-60	45-55	50-65
Fat	2-8	7-20	2-6	1.5-3.0
Ash	9-14	8-10	5-9.5	3-7
Nucleic acids	7-10	3-8	6-12	8-12

Bacterial protein is similar to fish protein, yeast's protein resembles soya and the fungi protein is somewhat lower than the yeast's. Of course microbiological proteins are deficient in the sulphur amino acids cysteine and methionine and require supplementation, while they exhibit better levels of lysine. The vitamins of micro-organisms are primarily of the B type, B12 occurs mostly in bacteria, while vitamin A is usually found in algae.

Microorganisms Used for SCP Production

Microorganism	Substrate	Used as	Used commercially
Algae			
Chlorella sp.	CO_2 + sunlight	Feed	Japan and Taiwan
Scenedesmus acutus	CO_2 + sunlight	-	
Spirulina maxima	CO_2 + sunlight	Feed	Mexico
Yeasts			
Candida utilis (Torula Yeast)	1. Confectionery	-	U.K.; Symba process
	2. Ethanol	Feed	USA
	3. Sulphite liquor	-	Europe, USA, Russia
C. intermedia	Whey	-	Vienna process
C. krusei (+ *Lactobacillus bulgarius*)	Whey	-	Kiel process
C. lipolytica	n-alkanes (C_{10} - C_{23}) + ammonia	-	Russia
Kluyveromyces fragilis	Whey	Food	France; Le Bel process
Saccharomyces cerevisiae	Molasses	(Food)*	Yes
Fungi			
Chaetomium cellulolyticum	Cellulosic wastes	-	Promising
Fusarium graminearum	Starch hydrolysate		U.K
Paecilomyces varioti	Sulphite liquor	-	Finland; Pekilo process
Bacteria			
Brevibacterium sp.	C1 - C4 hydrocarbons	-	Process developed
Methylophilus methylotrophus *Methylomonas* *Methylococcus*	Methanol	Feed	U.K.

Substrates for SCP production

A variety of substrates are used for SCP production. However, availability of necessary substrates is of considerable biological and economic importance for the production of SCP. Algae which contain chlorophylls, do not require organic wastes. They use free energy from sunlight and carbondioxide from air, while bacteria (except photoautotrops) and fungi require organic wastes, as they do not contain chlorophylls. Until recently genetic engineering has had a little economic impact on SCP synthesis.

(i) SCP from Petroleum based hydrocarbons

British Petroleum (BP) uses mutants of two yeasts, *Candida lipolytica* and *C. tropicalis* for fermenting n-alkanes which are petrochemical byproducts and produce SCP with high protein and also some vitamins. The product produced was called "TOPRINA"

(ii) SCP from Methanol

At Imperial Chemicals Industries (ICI), United Kingdom the genetic make up of *Methylophilus methylotrophus* has been altered so that it can grow readily on Methanol. The genetic alteration was accomplished by transferring a gene from *E.coli* to *M. methylotrophus.* The product was named "PRUTEEN". Pruteen contained 72% crude protein and was marketed for feed as a source of energy, vitamins and minerals as well as a highly balanced protein source. Unfortunately Pruteen now cannot compete with soya and fish meal. So today Pruteen, although a major engineering success is not economical to run.

(iii) SCP from Methane

Methane is cheap, abundant and without the toxicity problems of alkanes. The general *Methylomonas* and *Methylococcus* have been recognised as utilising methane as a carbon source. The species which has been extensively studied is *Methylomonas methanica.*

(iv) Mycoprotein

This is the only mycoprotein (except edible mushrooms) that has been cleared for human consumption. It uses a *Fusarium graminearum* growing in molasse, or glucose. The medium contains NH_3 for nitrogen source and pH control. The product is heat treated for RNA reduction. The mycelium is separated by vacuum filtration.

(v) SCP from Lignocellulose

The lignocellulosic wastes, mainly from agriculture, constitute the most abundant substrate for SCP which is also renewable. The world annual production of straw for example reaches 600 million tons every year. For the utilisation of lignocellulose, a pre-treatment is usually necessary. Many pre-treatment methods have been reported which vary from alkali or acid treatment, steam explotion or even x-ray radiation. To the present time the only economical utilisation of lignocellulosic wastes is in mushroom production . Besides our well known cultivated mushroom *Agaricus bisporus* there are many important ones which contain lignocellulolytic enzymes and are cultivated for food mainly in Asia and Africa. Some are of great economic significance and are cultivated on an industrial scale. Examples of important ones include the following species: *Volvariella* sp., *Lentinus edodes* and *Pleurotus* sp.

(vi) Carbohydrates

Forests are the rich source of carbohydrate in the form of cellulose but for using this substrate in SCP industries first they have to be converted to glucose by cellulose degrading microbes. Recently, high cellulase producing mutant fungi are being developed.

Applications / Uses

1. Good supplement of protein in human diet since some high yielding SCPs provide protein to the tune of even 72%
2. Single cell protein seems to be utilized better by ruminant animals
3. Suitable supplemental protein source for lactating dairy goats. Milk production and milk production efficiency was increased when single cell protein replaced groundnut meal in lactating goat diets
4. In poultry, Layers performance was optimized when single cell protein was fed at 2.5% of dry matter or 10% of dietary protein
5. Single cell protein was found to be able to replace up to 55% of the fish meal and soybean meal in swine diets
6. When single cell protein was included in prawn diets there was an increase in gains and feed conversion.

Problems in SCP production

1. Producing and harvesting microbial proteins involves high costs. In nearly all instances where a high rate of production would be achieved, the single cell protein will be found in rather dilute solutions, usually less than 5% solids. Methods available for concentrating include, filtration, precipitation, coagulation, centrifugation, and the use of semi-permeable membranes.
2. The de-watering methods require equipment that is quite expensive and would not be suitable for most small-scale operations. Removal of the amount of water necessary to stabilize the material for storage, in most instances, is not currently economical.
3. Single cell protein must be dried to about 10% moisture, or condensed and acidified to prevent spoilage from occurring, or fed shortly after being produced.

Spirulina as SCP

Spirulina maxima was sold in the name 'tecuitlatl' in 1821 by Bernal Diaz del Castillo in Mexico. In 1964, J. Leonard noticed dried biscuits of blue green color consisted of *Spirulina platensis* in several village markets. Many pilot plants for the production of *Spirulina* powder have been established in Japan, United States and European countries. Sosa Texcoco is the first Mexican Company to set up the first pilot plant in 1973 which produced about 150 tonnes *Spirulina* powder per annum; but the yield was increased to 1,000 tonnes in 1982.

Advantages of *Spirulina* over other algal SCP

Mass cultivation of *Spirulina* offers several advantages over *Chlorella* and *Scenedemus* as given below:

1. Being a filamentous alga, *Spirulina* can be harvested by simple and less expensive methods such as nylon or cotton cloth filter.
2. Filaments of *Spirulina* float on water surface due to presence of gas vacuoles. Hence, there is no problem of harvesting unlike *Chlorella* and *Scenedesmus.*
3. There is least chance of contamination in growth tanks of *Spirulina* as it grows at high alkaline pH 8-11.
4. Heat drying is sufficient for *Spirulina* as it has thin cell wall, whereas spray drying is required for *Chlorella* and *Scenedesmus* which is expensive,
5. No adverse effects on animals and humans
6. *Spirulina* is highly digestive (85-95%) due to thin wall and low nucleic acid contents (4%). It contains high percentage of digestible proteins (62-72%), vitamins, amino acids and other nutrients.

❑❑❑

Chapter 32

Microbiology of Food Canning

History of Food Canning

Nicolas Appert

The canning process dates back to the late 18th century in France when the Emperor Napoleon Bonaparte, concerned about keeping his armies fed, offered a cash prize to whoever could develop a reliable method of food preservation. *Nicolas Appert* conceived the idea of preserving food in bottles, like wine and realized if food is sufficiently heated and sealed in an airtight container, it will not spoil. Appert, is known as the *"father of canning,"*. An Englishman, *Peter Durand*, took the process one step farther and developed a method of sealing food into unbreakable tin containers, which was perfected by Bryan Dorkin and John Hall, who set up the first commercial canning factory in England in 1813.

What is canning?

Canning is an important, safe method for preserving food if practiced properly. The canning process involves placing foods in jars or similar containers and heating them to a temperature that destroys micro-organisms that cause food to spoil. During this heating process air is driven out of the jar and as it cools a vacuum seal is formed. This vacuum seal prevents air from getting back into the product bringing with it contaminating micro-organisms.

Safe Canning Methods

There are two safe ways of processing food, the boiling water bath method and the pressure canner method: The method to be followed depends on the type of food (low acid/high acid).

Boiling water bath method

In this method, jars of food are heated completely covered with boiling water (212°F at sea level) and cooked for a specified amount of time. The boiling water bath method is safe for high acid foods. The high acidic foods include:

- Fruits
- Properly pickled vegetables

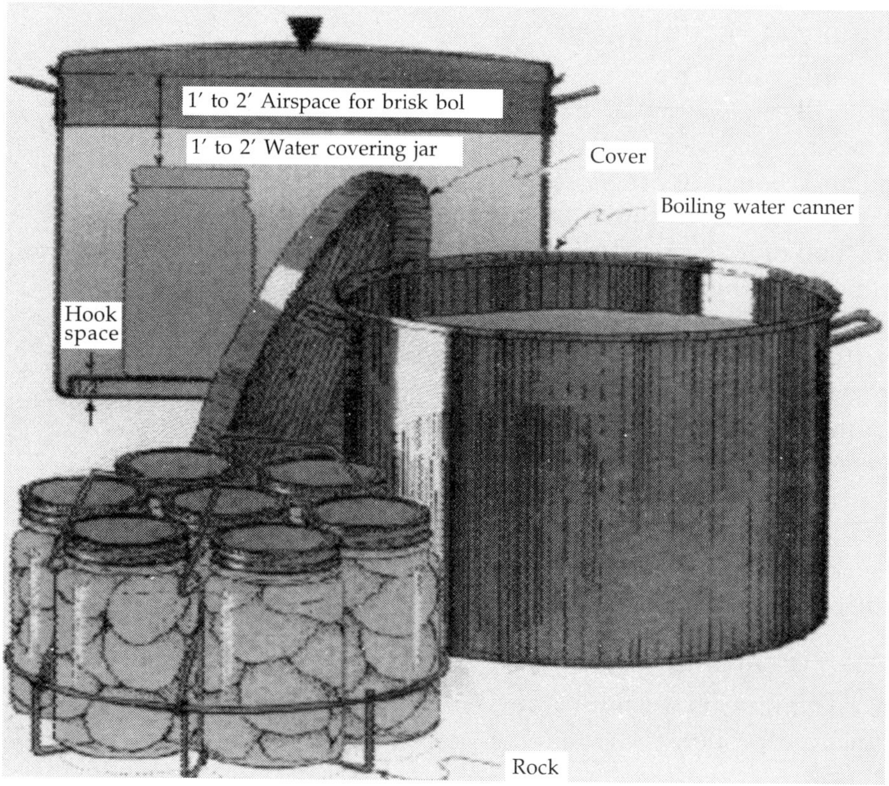

- Tomatoes
- Jams
- Jellies

High acid foods have a pH of 4.6 or less and contain enough acid so that the *Clostridium botulinum* spores can not grow and produce their deadly toxin. Certain foods like, tomatoes and figs, that have a pH value close to 4.6 need to have acid added to them in order to use the water bath method. This is accomplished by adding lemon juice of citric acid.

Steps Involved in water Bath method

- Fill the canner about halfway with hot water
- Turn on the burner and heat the water
- Allow the proper headspace so that all the extra air will be removed during processing, and a tight vacuum seal will be formed.
- Wipe off the rims of the jars with a clean, damp cloth.
- Screw on the lids, but not too tightly — air needs to escape during processing.
- Put glass jars filled with food stuff on the rack in the canner.
- Maintain water level at least 1 inch over the tops of the jars.
- Put the lid on the canner
- Boil gently and steadily for the recommended time, (generally 10 minutes) adjusting the heat and adding more boiling water as necessary.
- Use a jar lifter to carefully remove the jars as soon as the processing time is up.
- Do not tighten the lids. Allow the jars to cool untouched for 12 to 14 hours.

Pressure canner method: Jars of food are placed in 2 to 3 inches of water in a special pressure cooker which is heated to a temperature of at least 240° F in pressure canner method. Pressure canning is the only safe method of preserving low acid foods since they can be easily colonized by *Clostridium botulinum* whose cells are killed at boiling temperatures, they can form spores that can withstand these temperatures. The spores grow well in low acid foods, in the absence of air and produce the deadly botulinum toxins. The low acidic foods include:

- Meats
- Seafood
- Poultry
- Dairy products
- All vegetables

Steps Involved in Pressure Canner Method

- Place 2 to 3 inches of water in the canner. It should be hot but not boiling when canning raw-packed food; hot or gently boiling for hot-packed foods.
- Allow proper headspace, remove air bubbles, wipe jar rims and put on lids.
- Set the jars of food on the rack in the canner
- Fasten the canner lid & turn heat to high and watch until steam begins to escape from the vent. Let the steam escape steadily for 10 minutes.
- Let the canner heat quickly at first and adjust the burner temperature down slightly and let the pressure continue to rise to 15-18 pounds pressure.
- Keep the pressure constant by regulating the heat under the canner for 20 minutes.
- When the processing is completed, (generally 20 minutes) carefully remove the canner from the heat
- Do not tighten the lids. Allow the jars to cool, untouched for 12 to 24 hours

Diagrammatic representation of Steps Involved in Pressure Canner Method

Spoilage of canned food

The incidence of spoilage in canned foods is low, but when it occurs it must be investigated properly. Swollen cans often indicate a spoiled product. Microbial spoilage and hydrogen, produced by the interaction of acids in the food product with the metals of the can, are the principal causes of swelling. High summer temperatures and high altitudes may also increase the degree of swelling. Some microorganisms that grow in canned foods, however, do not produce gas and therefore cause no abnormal appearance of the can; nevertheless, they cause spoilage of the product.

Appearance of Spoiled Canned Foods

Spoiled canned food can be detected by observing the appearance of the cans themselves:

Flat - a can with both ends concave; it remains in this condition even when the can is brought down sharply on its end on a solid, flat surface.

Flipper - a can that normally appears flat; when brought down sharply on its end on a flat surface, one end flips out. When pressure is applied to this end, it flips in again and the can appears flat.

Springer - a can with one end permanently bulged. When sufficient pressure is applied to this end, it will flip in, but the other end will flip out.

Soft swell - a can bulged at both ends, but not so tightly that the ends can be pushed in somewhat with thumb pressure.

Hard swell - a can bulged at both ends, and so tightly that no indentation can be made with thumb pressure. A hard swell will generally "buckle" before the can bursts. Bursting usually occurs at the double seam over the side seam lap, or in the middle of the side seam.

During spoilage, cans may progress from normal to flipper, to springer, to soft swell, to hard swell. However, spoilage is not the only cause of abnormal cans. Overfilling, buckling, denting, or closing while cool may also be responsible.

Types of spoilage based on spoilage organism and symptoms

Type of spoilage	Symptoms	Spoilage organism
Thermophilic spoilage		
Flat sour spoilage	production of acid and no gas, food becomes sour, can remains flat	*Bacillus* spp. such as *B. coagulans* and *B. stearothermophilus*
Thermophilic anaerobic (TA) spoilage	can swells and may burst due to production of CO_2 and H_2, food becomes fermented sour, cheesy, and develops butyric odour	*Clostridium thermosaccharolyticum*
Sulfide spoilage	Production of H_2S gas, blackened food, rotten egg smell	*Clostridium nigricans*
Putrefactive anaerobic spoilage	can swells and may burst, putrid odour	*Clostridium sporogenes*
Aerobic spore former's spoilage	If the canned food is cured meat, swelling is seen	*Bacillus* spp
Mesophilic spoilage		
Butyric acid fermentation	Can swells due to the production of CO_2 and H_2, fermented butyric odor	*Clostridium butyricum* and *C. pasteurianum*
	acidic (tomato juice, fruits, fruit juices, etc.) or medium acidic (corn, peas, spinach, etc.) foods	
Lactic acid fermentation	Can swells due to the production of CO_2 , Acid odor canned sea-foods, meats	*Streptococcus* sp., *Pseudomonas*, *Proteus*, Yeasts

Guidelines for proper storage of Canned Food

- Store in a cool, clean, dry place where temperatures are below 85 degrees (between 50-70 degrees is good) but not freezing temperatures.
- Rotate foods so the oldest is used first. Try not to keep canned foods more than one year.
- Use canned meats and seafood within 12 months.
- Use low-acid canned foods within 8-12 months.

- Use high-acid foods within 12 to 18 months. Foods stored longer will be safe to eat if the cans show no signs of spoilage or damage but may deteriorate in color, flavor and nutritive value.
- Canned fruit juices can be stored and used up to 3 years.

❑❑❑

Chapter 33

Biofuels : Alcohols Hydrogen and Methane

Biofuel

Biofuels are produced from living organisms or from metabolic by-products (organic or food waste products). In order to be considered as biofuel the fuel must contain over 80 percent renewable materials.

I. Alcohols as biofuel

Production of ethyl alcohol from sugary materials is one of the oldest known microbiological process. Alcohol is an important solvent and raw material used in a variety of chemical industries. Although today industrial alcohol is also produced synthetically from ethylene, production of alcohol by fermentation of cheap sugary materials such as molasses by yeast is still an important industry.

For ethyl alcohol production, selected strains of *Saccharomyces cerevisae* are employed since all the strains are not equally efficient. The alcohol tolerance and sugar tolerance are important criteria used in the selection of yeast strains. Strains tolerant to high sugar and alcohol concentration are desired. The raw material generally used is either crude cane molasses or best molasses which contain about 50 per cent fermentable sugars. The production process involves the dilution of molasses to a suitable sugar concentration (15- 16 per cent sugars), addition of small quantity of nitrogen source (urea, ammonium sulphate or ammonium phosphate), adjustment of pH to about 5.0 and the addition of an actively growing yeast culture. The fermentation is carried out in big deep tanks of steel or stainless steel. The fermentation is allowed to continue for about 24 – 36 h at 25°C to 30°C after which the cells are allowed to settle. The fermented mash is then distilled and passed through rectifying columns to recover ethyl alcohol. A large amount of carbon-di oxide is also produced during the fermentation which is purified and compressed.

The yield of ethyl alcohol is about 50 per cent of the fermentable sugar concentration. Further purification of ethyl alcohol is done by fractional distillation. In some distilleries, the yeast is recovered and used as animal feed while in most, it is discarded into the effluents, a procedure that is very undesirable.

In recent years because of the possibility of using ethyl alcohol as a fuel supplement and a chemical feed stock, there is increased interest in increasing production but at a cheaper and economical rate. For this, a variety of improvements in the traditional batch fermentation have been described in literature. Among these, the one that has attracted attention is the cell recycle technique which does not involve much additional expenditure. Basically, the technique involves the reuse of cell mass that is produced during the fermentation. It has been found that by doing so, about 5-10 per cent of the substrate which would have been otherwise used for cell growth is saved in addition to a great saving in the cost of inoculum and time. By using recycling technology, fermentation time has been drastically reduced from 24-36 hours in a batch fermentation to as low as 5-6 hours. (production techniques are already discussed in chapter 27).

Distilled Alcoholic Beverages

Whisky, rum, gin, brandy and vodka, are all products of alcoholic fermentation of different raw materials such as grains, molasses, potatoes and fruits. They are distilled products containing higher content of alcohol than either wine or beer. In principle, production of these alcoholic beverages is similar to the production of industrial alcohol. The quality of the alcohol produced and used for the production of these beverages depends on the type of secondary fractional alcohol distillation. It is essential to derive a pure spirit for the production of these beverages.

Whisky is traditionally produced from the fermentation of grain mash, rum from molasses, brandy from grape wines and vodka from potatoes. The flavour and aroma are introduced into the purified alcohol during the blending process. Sometimes, these alcoholic beverages are allowed to age for long periods in wooden casks before they are consumed.

II. Methane gas as biofuel

Methane fermentation is a versatile biotechnology capable of converting almost all types of polymeric materials to methane and carbon dioxide under anaerobic conditions. This is achieved as a result of the consecutive biochemical breakdown of polymers to methane and carbon dioxide in an environment in which varieties of microorganisms harmoniously grow and produce reduced end-products. Anaerobes play important roles in establishing a stable environment at various stages of methane fermentation.

Stages of methane formation

Methane fermentation is the consequence of a series of metabolic interactions among various groups of microorganisms. The first group of microorganisms secretes enzymes which hydrolyze polymeric materials to monomers such as glucose and amino acids, which are subsequently converted to higher volatile fatty acids, H_2 and acetic acid. In the second stage, hydrogen-producing acetogenic bacteria convert the higher volatile fatty acids e.g., propionic and butyric acids, to H_2, CO_2, and acetic acid. Finally, the third group, methanogenic bacteria convert H_2, CO_2, and acetate, to CH_4 and CO_2.

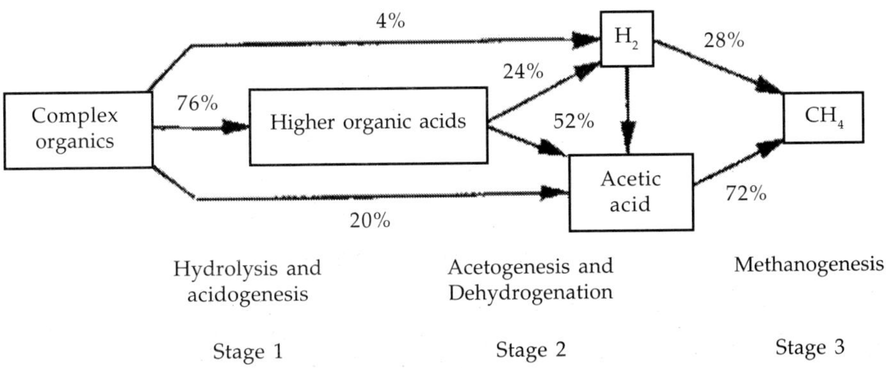

Stage 1: Hydrolysis and acidogenesis

Polymeric materials such as lipids, proteins, and carbohydrates are primarily hydrolyzed by extracellular, hydrolases, excreted by microbes present in Stage 1. Hydrolytic enzymes, (lipases, proteases, cellulases, amylases, etc.) hydrolyze their respective polymers into smaller molecules, primarily monomeric units, which are then consumed by microbes.

1. Lipases convert lipids to long-chain fatty acids. A population density of 10^4 - 10^5 lipolytic bacteria per ml of digester fluid has been reported. Clostridia and the micrococci appear to be responsible for most of the extracellular lipase producers. The long-chain fatty acids produced are further degraded by p-oxidation to produce acetyl CoA.
2. Proteins are generally hydrolyzed to amino acids by proteases, secreted by *Bacteroides, Butyrivibrio, Clostridium, Fusobacterium, Selenomonas,* and *Streptococcus.* The amino acids produced are then degraded to fatty acids such as acetate, propionate, and butyrate, and to ammonia as found in *Clostridium, Streptococcus, Selenomonas, Campylobacter,* and *Bacteroides.*

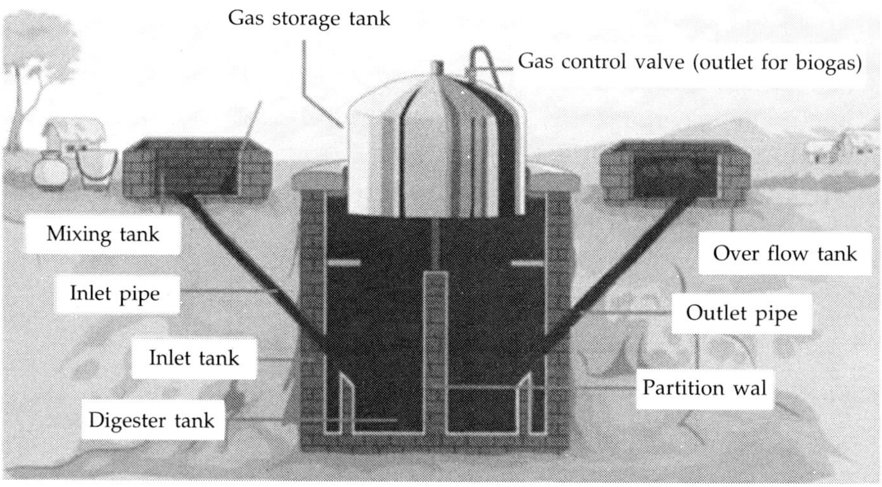

Fixed Dome type Biogas Plant

3. Polysaccharides such as cellulose, starch, and pectin are hydrolyzed by cellulases (endo-β-1,4-glucanases, exo-β-1,4-glucanases, cellobiase), amylases (a-amylases, β-amylases, amyloglucosidases, maltase) and pectinases.

4. Pectins are degraded by pectinases, including pectin esterases and depolymerases. Xylans are degraded with endo-xylanase and xylosidase to produce xylose.

5. Hexoses and pentoses are generally converted to C_2 and C_3 intermediates and to reduced electron carriers (e.g., NADH) via common pathways. Most anaerobic bacteria undergo hexose metabolism via the Emden-Meyerhof-Parnas pathway (EMP) which produces pyruvate as an intermediate along with NADH. The pyruvate and NADH thus generated, are transformed into fermentation end products such as lactate, propionate, acetate, and ethanol by other enzymatic activities which vary tremendously with microbial species.

Thus, in hydrolysis and acidogenesis (Stage 1), sugars, amino acids, and fatty acids produced by microbial degradation of biopolymers are successively metabolised by fermentation end-products such as lactate, propionate, acetate, and ethanol by other enzymatic activities which vary tremendously with microbial species.

Stage 2: Acetogenesis and dehydrogenation

Although some acetate (20%) and H_2 (4%) are directly produced by acidogenic fermentation of sugars, and amino acids, both products are primarily derived from the acetogenesis and dehydrogenation of higher volatile fatty acids (Stage 2).

Obligate H_2-producing acetogenic bacteria are capable of producing acetate and H_2 from higher fatty acids. Only *Syntrophobacter wolinii,* a propionate decomposer and *Sytrophomonos wolfei,* a butyrate decomposer have thus far been isolated due to technical difficulties involved in the isolation of pure strains, since H_2 produced, severely inhibits the growth of these strains. The use of co-culture techniques incorporating H_2 consumers such as methanogens and sulfate-reducing bacteria may therefore facilitate elucidation of the biochemical breakdown of fatty acids.

Stage 3: Methanogenesis

Methanogens are physiologically united as methane producers in anaerobic digestion (Stage 3). Although acetate and H_2/CO_2 are the main substrates available in the natural environment, formate, methanol, methylamines, and CO are also converted to methane (CH_4)

Since methanogens, as obligate anaerobes, require a redox potential of less than -300 mV for growth, their isolation and cultivation was somewhat elusive due to technical difficulties encountered in handling them under completely O_2-free conditions. However, as a result of a greatly improved methanogen isolation techniques developed by Hungate, more than 40 strains of pure methanogens have now been isolated. Methanogens can be divided into two groups: H_2/CO_2- consumers and acetate-consumers. Although some of the H_2/CO_2-consumers are capable of utilizing formate, acetate is consumed by a limited number of strains, such as *Methanosarcina* spp. and *Methanothrix* spp. (now, *Methanosaeta),* which are incapable of using formate. Since a large quantity of acetate is produced in the natural environment, *Methanosarcina* and *Methanosaeta* play an important role in completion of anaerobic digestion and in accumulating H_2, which inhibits acetogens and methanogens. H_2-consuming methanogens are also important in maintaining low levels of atmospheric H_2. H_2/CO_2-consuming methanogens reduce CO_2 as an electron acceptor via the formyl, methenyl, and methyl levels through association with unusual coenzymes, to finally produce CH_4.

III. Hydrogen gas as Biofuel

Hydrogen gas is seen as a future energy carrier by virtue of the fact that it is renewable, does not evolve the "greenhouse gas" CO_2 in combustion, liberates large amounts of energy per unit weight in combustion, and is easily converted to electricity by fuel cells. Biological hydrogen production has several advantages over hydrogen production by photoelectrochemical or thermochemical processes. Biological hydrogen production by photosynthetic microorganisms for example, requires the use of a simple solar reactor such as a transparent closed box, with low energy requirements. Electrochemical hydrogen production via solar battery-based water splitting on the other hand requires the use of solar batteries with high energy requirements.

a. Hydrogen from Biophotolysis of water

Microalgae are primitive microscopic plants living in aqueous environments. Cyanobacteria, formerly known as blue-green algae, are now recognized as bacteria since the anatomical characteristics of their cells are prokaryotic (bacterial type). Miroalgae and Cyanobacteria along with higher plants, are capable of oxygenic photosynthesis according to the following reaction:

$$CO_2 + H_2O \dashrightarrow CH_2O + O_2$$

Photosynthesis consists of two processes: light energy conversion to biochemical energy by a photochemical reaction, and CO_2 reduction to organic compounds such as sugar phosphates, through the use of this biochemical energy by Calvin-cycle enzymes. Under certain conditions, however, instead of reducing CO_2, a few groups of microalgae and Cyanobacteria consume biochemical energy to produce molecular hydrogen. Hydrogenase and nitrogenase enzymes are both capable of hydrogen production.

b. Hydrogenase-dependent hydrogen production

A green alga, *Scenedesmus,* produced molecular hydrogen under light conditions after being kept under anaerobic and dark conditions.

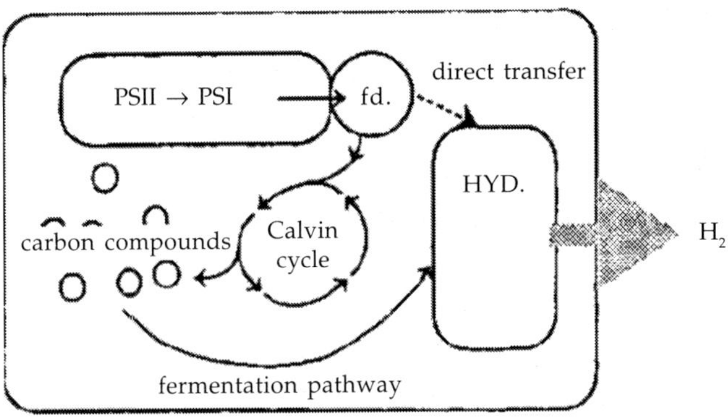

Hydrogenase-mediated hydrogen production

Hydrogenase, the enzyme responsible for this hydrogen production, catalyses the following reaction:

$$2H^+ + 2X_{reduced} \dashrightarrow 6\ H_2 + 2X_{oxidized}$$

The electron carrier, X, is thought to be ferredoxin. Since ferredoxin is reduced with water as an electron donor by the photochemical reaction, green alga are theoretically water-splitting microorganisms.

The reducing power (electron donation) of hydrogenase does not always come from water, but may sometimes originate intracellularly from organic compounds such as starch. The contribution of the decomposition of organic compounds to hydrogen production, is dependent on the algal species concerned, and on culture conditions. Even when organic compounds are involved in hydrogen production, an electron source can be derived from water, since organic compounds are synthesized by oxygenic photosynthesis. The reason for hydrogenase inactivity in green algae under normal photosynthetic growth conditions is unclear. Hydrogenase is thought to become active in order to excrete excess reducing power under specific conditions, such as anaerobic conditions.

Hydrogenase is too oxygen-labile for sustainable hydrogen production: light-dependent hydrogen production ceases within a few to several tens of minutes since photosynthetically produced oxygen inhibits or inactivates hydrogenases. A low oxygen concentration within the reaction vessel has to be maintained. A "short circuit" of photosynthesis, whereby hydrogen production and CO_2 fixation occurred by a single photosystem (photosystem II only) of a *Chlamydomonas* mutant Green algae are applicable in another method of hydrogen production.

Cyanobacteria also produce hydrogen gas auto-fermentatively under dark and anaerobic conditions. *Spirulina* species were demonstrated to have the highest activity among cyanobacteria tested. The nature of the electron carrier for hydrogenase in cyanobacteria is still unclear.

f

c. Nitrogenase-dependent hydrogen production

A nitrogen-fixing cyanobacterium is capable of producing hydrogen and oxygen gas simultaneously in an argon atmosphere for several hours. Nitrogenase is responsible for nitrogen-fixation and is distributed mainly among prokaryotes, including cyanobacteria, but does not occur in eukaryotes, under which microalgae are classified. Molecular nitrogen is reduced to ammonium with consumption of reducing power (e′ mediated by ferredoxin) and ATP. The reaction is substantially irreversible and produces ammonia:

$$N_2 + 6H^+ + 6e^- \text{------}\rightarrow 2HN_3$$

12ATP *f* 12(ADP+Pi)

However, nitrogenase catalyzes proton reduction in the absence of nitrogen gas (*i.e.* in an argon atmosphere).

$$2H^+ + 2e^- \text{------}\rightarrow H_2$$

4ATP 4(ADP+Pi)

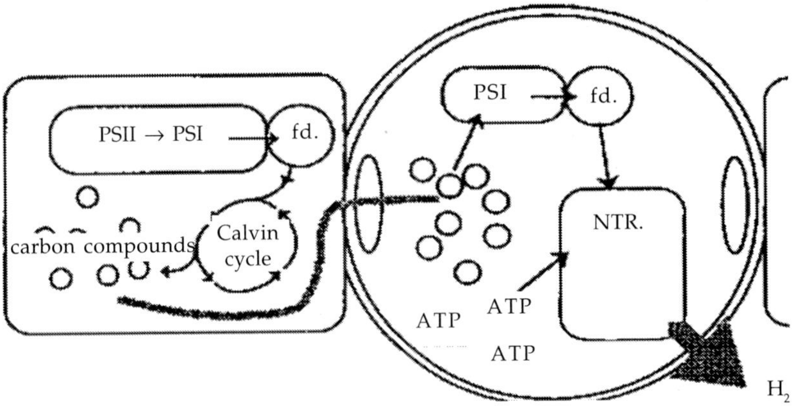

Nitrogenase-mediated hydrogen production in heterocystous cyanobacteria

Hydrogen production catalyzed by nitrogenase occurs as a side reaction at a rate of one-third to one-fourth that of nitrogen-fixation, even in a 100% nitrogen gas atmosphere.

Nitrogenase itself is extremely oxygen-labile. Unlike in the case of hydrogenase, however, cyanobacteria have developed mechanisms for protecting nitrogenase from oxygen gas and supplying it with energy (ATP) and reducing power. The most successful mechanism is the localization of nitrogenase in the heterocysts of filamentous cyanobacteria. Vegetative cells (ordinary cells) in filamentous cyanobacteria carry out oxygenic photosynthesis. Organic compounds produced by CO_2 reduction are transferred into heterocysts and are decomposed to provide nitrogenase with reducing power. ATP can be provided by PSI-dependent and anoxygenic photosynthesis within heterocysts. Investigations into prolongation and optimization of hydrogen production, revealed that the hydrogen-producing activity of cyanobacteria was stimulated by nitrogen starvation.

The presence and physiological roles of hydrogenases in nitrogen-fixing cyanobacteria remains controversial, but 'uptake' hydrogenase appears to consume and re-use hydrogen gas, resulting in a decrease in net hydrogen production. Aerobic hydrogen production by a nitrogen-fixing *Anabaena* sp., believed to be an uptake hydrogenase-deficient strain.

❑❑❑

Chapter 34

Microbiology of Potable Water

Drinking water is obtained either from surface sources such as rivers, lakes or from underground water. Such natural waters are likely to be polluted. Although water purification systems envisage protection from pollution, sometimes, the water supply can become a potential carrier of pathogenic organisms and endanger public health. A number of diseases such as cholera, typhoid, viral hepatitis *etc.*, are known to be water borne. These pathogens are commonly transmitted through drinking water and cause infection of the intestinal tract. It is therefore, necessary to employ treatment facilities to purify water and to provide safe drinking water (Potable water).

What is Potable water?

Water that is free from diseases producing organism and chemical substances deleterious to health is called potable water.

Purification process

The main operations employed in water purification to produce potable water are : (i) sedimentation, (ii) filtration, and (iii) chlorination. Most microorganisms are removed during coagulation with aluminium sulphate and sand filtration and subsequent treatment of water with chlorine (0.2 – 1 mg free chlorine per liter) will ensure its portability.

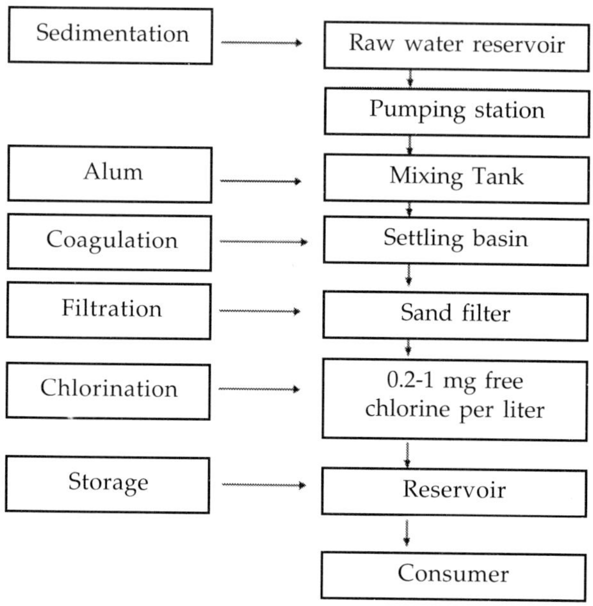

Main operations in drinking water purification

Microbiological Quality

Water can be perfectly clear in appearance and free from odour and taste and yet, be contaminated by microorganisms. Pathogenic organisms enter into water through sewage contamination or discharges from animals or humans into the reservoirs. The coliforms (E.coli and related organisms), *Streptococcus faecalis* and *Clostridium perfringens* which are normal inhabitants of the large intestine of animals and humans enter water supplies through faecal contamination. The presence of any of these bacterial species in water is evidence of sewage or faecal pollution. Techniques are available by which the presence of these specific groups can be easily identified. The routine bacteriological examination consists of (i) Plate count to determine the number of bacteria present, and (ii) biochemical test to reveal the presence of coliform bacteria. A variety of other bacteria and organisms which may not be serious pathogens including faecal streptococci, slime forming bacteria, Sulphur bacteria, algae etc. may also cause problems of odour, color and taste and it is essential that be eliminated from the drinking water.

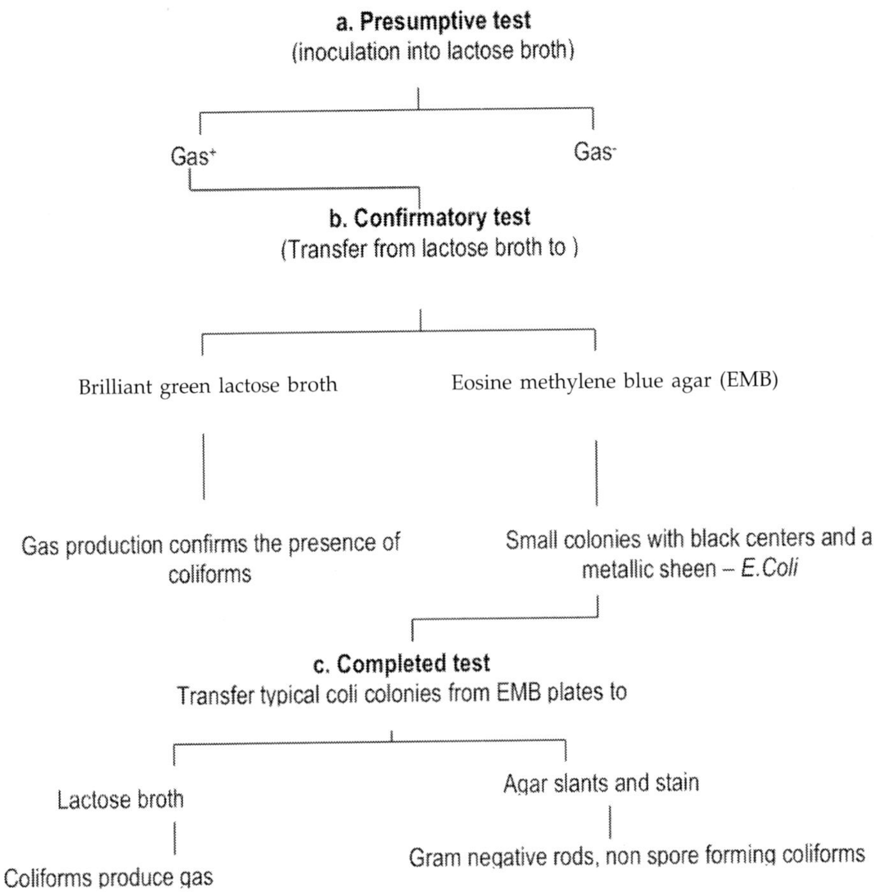

Laboratory tests for detecting contamination by coliforms

International standards for drinking water

a. Fluoride: For drinking purposes, water should be tested for fluoride ion concentration. Its deficiency in drinking water is harmful to humans and causes health issues such as tooth decay etc. In drinking water Soluble fluoride concentration is up to 1 ppm or 1 $mgdm^{-}3$. However, F^{-} ion concentration above 2 ppm causes brown mottling of teeth. At the same time, excess fluoride (over 10 ppm) causes harmful effect to bones and teeth

b. Lead: Drinking water gets contaminated with lead when lead pipes are used for transportation of water. The prescribed upper limit concentration of lead in drinking water is about 50 ppb. Lead can damage kidney, liver, reproductive system etc.

c. *Sulphate*: Excessive sulphate (>500ppm) in drinking water causes laxative effect, otherwise at moderate levels it is harmless.

d. *Nitrate*: The maximum limit of nitrate in drinking water is 50 ppm. Excess nitrate in drinking water can cause disease such as methemoglobinemia ('blue baby' syndrome).

❒❒❒

Selected References

1. Black, J.G. (1999). Microbiology : Principles and Explorations. Prentice Hall Inc., New Jersy.

2. Casida, L.E.J.R. (1968). Indusrial Microbiology. New Age Publishers

3. Joanne M.Willey, Linda M.Sherwood and Christopher J.Woolverton. (2008). Prescott, Harley and Klein's Microbiology. McGraw-Hill Publishers. New York.

4. Larry R.B., Thomas J.M., Michael P.D. (2001). Food Microbiology: Fundamentals and Frontiers. ASM Press.

5. Madigan, M.T., Martinko J.M. and Parker J. (2003). Brock Biology of Microorganisms. Pearson Education Inc., New Jersy.

6. Pelczar, Jr. M.J., Chan, E.C.S. and Krieg, N.R. (2010). Microbiology- An Application Based Approach. Tata McGraw Hill Education Pvt. Ltd., New Delhi

7. Subba Rao, N.S. (1999). Soil Microbiology. Soil Microorganisms and Plant Growth. Oxford & IBH Publishing Co., New Delhi